THE MOLECULE AND ITS DOUBLE

THE MCGRAW-HILL HORIZONS OF SCIENCE SERIES

The Science of Crystals, Françoise Balibar

Water, Paul Caro

Universal Constants in Physics, Gilles Cohen-Tannoudji

The Dawn of Meaning, Boris Cyrulnik

The Realm of Molecules, Raymond Daudel

The Power of Mathematics, Moshé Flato

The Gene Civilization, François Gros

Life in the Universe, Jean Heidmann

Our Changing Climate, Robert Kandel

The Language of the Cell, Claude Kordon

The Chemistry of Life, Martin Olomucki

The Future of the Sun, Jean-Claude Pecker

How the Brain Evolved, Alain Prochiantz

Our Expanding Universe, Evry Schatzman

The Message of Fossils, Pascal Tassy

Earthquake Prediction, Haroun Tazieff

JEAN

JACQUES

THE MOLECULE AND ITS DOUBLE

McGraw-Hill, Inc.

New York St. Louis San Francisco Auckland Bogotá
Caracas Lisbon London Madrid Mexico
Milan Montreal New Delhi Paris
San Juan São Paulo Singapore
Sydney Tokyo Toronto

English Language Edition

Translated by Lee Scanlon
in collaboration with
The Language Service, Inc.
Poughkeepsie, New York

Typography by AB Typesetting
Poughkeepsie, New York

Library of Congress Cataloging-in-Publication Data
Jacques, Jean.
[*Molécule et son double*. English]
The Molecule and its Double / Jean Jacques.
p. cm. — (The McGraw-Hill *Horizons of Science* series)
Translation of: *La Molécule et son double*.
Includes bibliographical references.
ISBN 0-07-032399-2
1. Stereochemistry. 2. Chirality. I. Title. II. Series.
QD481.M2713 1993 93-3309
541.2'23 — dc20

Copyright © 1993 by McGraw-Hill, Inc. All rights reserved. Except as permitted under the United States Copyright Act of 1976, no part of this publication may be reproduced or distributed in any form or by any means, or stored in a database or retrieval system, without the prior written permission of the publisher.

The original French language edition of this book
was published as *La Molécule et son double*, copyright © 1992,
Hachette, Paris, France.
Questions de science series
Series editor, Dominique Lecourt

This book is printed on recycled, acid-free paper containing a minimum of 50% recycled de-inked fiber.

TABLE OF CONTENTS

INTRODUCTION

In a treatise entitled *Prolegomena zur einer jeden künftigen Metaphysik die als Wissenschaft wird auftreten können* [Prolegomena to any future Metaphysics], published in 1783, the philosopher Emmanuel Kant (1724-1804) posed a question which half a century later was to be at the heart of one of the most irritating and at the same time one of the most productive enigmas that modern science has had to deal with: "What could be more similar to my hand or my ear, and more equal to it in every respect, than its mirror image? And yet I cannot substitute the hand as seen in the mirror for its original; for if it is a right hand, the other in the mirror is a left hand, and the image of the right ear is a left ear, which cannot take the place of the other either." The author of the *Critique of Pure Reason* (1781) saw in this a confirmation of that work's key thesis, the very same thesis that was to determine the fate of philosophy for more than a century: The hand as perceived by us in space comes under the concept of the hand as we form it with our understanding. Yet it is not reducible to such a concept, since it actually exists in our world only as "right" or "left." Space introduces a difference "between two similar and equal things." We never know things in themselves, but only as they present themselves to our understanding, already ordered by our "sensibility," which disposes them, first of all, in space—which, Kant explains, is not to be sought either in things or between them, but is imposed on reality by the structure of our mind.

We are not concerned here with the philosophical implications that Kant saw in this thesis; it will suffice to note that he thus called attention to properties of symmetry which up to that time, paradoxically, had given rise to very little study on the part of geometricians.

In the early 19th century, chemists, on an altogether different scale, came up against the same type of question when they groped their way along in the investigation of the molecular structure of matter: They discovered objects that were identical in every respect, but proved not to be superimposable. It was not long before they spontaneously rediscovered the Kantian example of the hand, and Lord Kelvin (1824-1907) paid homage, no doubt unintentionally, to the German philosopher when he baptized this property "chirality" (from the Greek *kheir*, hand).

The adventure apparently started by surprise, and it seems to have been touch-and-go all the way. Louis Pasteur (1822-1895), the young and brilliant chemist, who had graduated from the *École Normale Supérieure* just two years before, showed in 1848 that certain molecules, and specifically those of the strange-acting racemic acid, are composed of a combination of two twin molecules, one of which proves to be the nonsuperimposable mirror image of the other. He managed to split what was apparently a single substance into a "right-handed" molecule and a "left-handed" molecule, which are identical and indistinguishable except for the spatial aspect. History has it that one of his teachers, the illustrious physicist Jean-Baptiste Biot (1774-1862), after asking him to repeat the experiment in his presence, came out with these words of kind admiration: "My dear child, I have so loved science in my life that I am utterly thrilled by this." Indeed, the importance of this discovery did not reside simply in the extreme experimental dexterity of a researcher who had only a few rudimentary instruments to work with: a pair of tweezers and a magnifying glass, a goniometer for measuring the angles of a crystal and a polarimeter for measuring its "rotatory power," i.e., the capacity of such a molecule for causing a beam of polarized light to rotate to the right or the left.

Jean Jacques, the well-known chemist, historian of his discipline and militant advocate of scientific popularization, relates the surprising developments and paradoxes of the history that started with this feat. Studies extending, rectifying and developing Pasteur's work on "molecular dissymmetry" were to give rise, 25 years later, under the signature of the Dutchman Jacobus Hendricus van't Hoff (1852-1911), to *Chemistry in Space*, today known as stereochemistry, invented simultaneously by the Frenchman Joseph-Achille Le Bel (1847-1930); even inorganic chemistry in turn underwent an upheaval, as a result of the research of the Swiss chemist Alfred Werner (1866-1919). In the lesson taught at the Chemical Society of Paris on January 20, 1860, Pasteur declared: "When I began to engage in my own work, I sought to strengthen myself in the study of crystals, since I foresaw how helpful it would be for my chemical research." The success achieved thanks to that "foresight" determined the course of the work he subsequently undertook, the work that brought him fame. Pasteur's studies on fermentation were directly in line with this early work, though mention is rarely made of this fact. In retrospect, the 1848 experiment can be seen as the starting point of a revolution in physiology as well as medicine.

However, if we look back today at the presuppositions and the consequences of that experiment from the philosophical standpoint, it proves even more exciting. We can readily see the vanity of epistemological discourses on the supposedly timeless rules of any "logic of scientific discovery." Where in fact did the young Pasteur get the idea, original at the time, that in order to do "chemical research" he first had to "strengthen" himself in crystallography? From the still quite recent intersection of several avenues of research that had long been unrelated.

The first was a continuation of the age-old controversies on the nature of light. It was to Étienne-Louis Malus (1775-1812) that Pasteur referred. Malus, a graduate of the École Polytechnique and veteran of

Napoleon's Egyptian campaign who was a friend of Pierre Simon de Laplace (1749-1827), had taken care to reproduce the experiments of Christiaan Huygens (1629-1695) on "double refraction." If a ray of light is passed through two calcite crystals, the intensity of the ray transmitted by the second crystal varies when the latter is turned, and the ray no longer splits in the second if it is oriented the same way as the first. This, according to Malus, proved Sir Isaac Newton (1642-1727) right and imposed a "molecular" conception of light. Light molecules were assumed to have "poles." When rays passed through calcite or were reflected, the poles became oriented: "Polarization," he said, took place. The word remained, but the concept changed. Augustin Fresnel (1788-1827), through his work on diffraction, had acquired the opposite conviction: Light was not "corpuscular," but "undulatory." Thus he, in turn, with François Arago (1786-1853), repeated Malus's experiments on polarization. Following James Young (1811-1883) in the study of the phenomenon of interference, they came up with quite a different interpretation, reaching the conclusion that "natural light" vibrates in all directions of the wave plane, whereas "rectilinearly polarized light" vibrates in only one direction of that plane.

The second line of research involved the work of those who had just delivered the crystal from the status of a "monster" or "curiosity" that it had maintained for centuries within the precincts of mineralogy. Jean-

Baptiste Romé de l'Isle (1736-1790) and the Abbé René Just Haüy (1743-1822) had not been satisfied with establishing its mineral nature against the tradition that still assigned it to the vegetable or animal kingdom. In geometrizing the crystal, they had turned it into a scientific object. The perfection of its surfaces quickly became less interesting than the irregular forms taken by its angles, as Françoise Balibar has so well pointed out in *The Science of Crystals*. Thus, in a certain sense, the second line of research, in order to intersect with the first, had to deprive it of its instrument, so as to turn it into its own object.

The third avenue of research had to do with the very heart of chemical theory. In 1823 Michel-Eugène Chevreul (1786-1889) had defined the chemical species as a "collection of entities identical in nature, proportion and arrangement of elements." As we shall see, however, a note by Eilhard Mitscherlich (1794-1863) came along in 1844 to threaten this definition, showing that two substances (sodium ammonium tartrate and paratartrate), though assigned to the same species thus defined, had different optical properties: One rotated plane polarized light, while the other did not. Claude Bernard (1813-1878) expressed in his notebooks this somewhat ironical judgment: "Pasteur sees only what he is aiming at." But at least he knew how to aim. Pasteur was in fact not the only one to be perplexed by Mitscherlich's note. He was, however, the only one who knew where to look next: into a minute, previously unnoticed difference in these two substances in the crystalline state.

In so doing, he introduced a new area of science. The goniometer, the mineralogist's instrument, measured the angles of the crystal, while the polarimeter was used in optics as an instrument for analyzing light. By manipulating the two, Pasteur opened up the study of the chemical properties of substances based on the spatial arrangement (symmetrical or asymmetrical) of their components as revealed by their optical properties. After this feat, Pasteur no longer concerned himself much with crystallography. He was able to consider himself sufficiently "strengthened" to take up his chemical studies. Yet he had made a discovery that determined what was to follow: The organic was distinguished from the inorganic by asymmetry. Everyone knows of the battle, inspired by that conviction, that he waged against the theory of "spontaneous generation" revived by Félix Pouchet (1800-1872). Pasteur put his discovery "to work" right away. In a series of experiments he found that molds destroyed only molecules oriented a certain way, leaving the others intact. He thus used the optical properties of a ferment, such as *Penicillium glaucum*, as a "biological analyzer" (Cl. Salomon-Bayet). This was the first step in what we call biochemistry.

We now know that one of the most remarkable properties of a living organism as such is its ability to take symmetrical molecular structures from its environment and use them to build carbon compounds that are asymmetric. A plant, for example, takes inorganic compounds such as water and carbon dioxide and uses them to synthesize asymmetric starches or sugars. That is not all: We know that the amino acids thanks to which our human organisms live and reproduce occur as "left-handed" asymmetrical molecules.

Yet this still does not fully account for the novelty or the impact of this research. It happened that the findings in this area were matched by questions that arose in a field of research that one might think was quite remote. These questions, too, were situated at the meeting point of two major scientific domains: particle physics and astrophysics. Evry Schatzman, in his work *Our Expanding Universe*, points out that if one goes back to a time when the density of radiant energy dominated the system that constituted the Universe, one observes that the number of photons per unit of volume was approximately the same as the number of all other particles. Today, however, the situation looks quite different. There is approximately one hydrogen nucleus per unit of volume for a quantity of photons ranging from 1 to 10 billion! How can we explain this destruction, this break in symmetry? How, moreover, can we explain that in our galaxy and in neighboring galaxies we deal only with matter and never with antimatter, with particles and never with antiparticles, though we know that they always "accompany" particles and we can easily produce them in our accelerators? Must we not view the characteristic dissymmetry of living things in relation to "cosmic dissymmetry?" The question becomes fascinating when we note that the electron itself tends to "turn to the left," and when we know that the binding energy of "left-handed" molecules is ever so slightly greater than that of "right-handed" molecules.

Pasteur would no doubt have been overjoyed by such findings, for his official positivism in no way prevented him from indulging in bold speculations: Unable to explain the origin of the dissymmetry of living molecules, he had imagined that it might be attributed to a cosmic phenomenon. Jean Jacques cites the extraordinary text in which the great chemist looks to terrestrial magnetism, still but little known at the time, for the key to this ultimate puzzle! Fifty years later, on September 8, 1898, a violent controversy arose in England in the journal *Nature*. In it, Professor Francis-Robert Japp published a lecture entitled "Stereochemistry and Vitalism," in which he called the attention of his colleagues to Pasteur's crystallographic studies. From the thesis that "only asymmetry engenders asymmetry" he drew the following conclusion: "The chance interaction of atoms, even if they had had eternity in order to meet and combine, could not have brought about the *tour de force* constituted by the formation of the first optically active organic compound. Coincidence is ruled out, and any purely mechanical explanation of the phenomenon is doomed to fail." He thus suggests that a supernatural force must be seen to have been at work here. Within two weeks, the celebrated biologist and statistician Karl Pearson (1857-1936) vigorously contested the validity of Japp's probabilistic argument. Herbert Spencer (1820-1903) himself, a favored philosopher of the United States, joined the fray, defending his agnostic positions, as did the Irish mathematician George Francis FitzGerald (1851-1901) and numerous lesser-known authors.

For his part, Louis Pasteur, a fervent Catholic, steadfastly refrained from any speculation of this kind. He criticized Pouchet's experiments in so far as people assumed that they proved "spontaneous generation," but he refused, whatever may have been said of him, to erect a definitive "wall" between the living and the nonliving. Moreover, ignoring the origin of chirality, he was in no hurry to convert chance into providence and continued, with whatever means were available, to look for rational links in order to come to understand the unknown. The lesson surely deserves to be studied today, at the threshold to the 21st century, as we witness a resurgence of Japp's arguments, flaming up afresh in connection with the cosmological scenario of the big bang theory in the guise of the famous "anthropic principle."

Dominique LECOURT

PROLOGUE

Beneath the appearances of the everyday world, chemistry tracks down the invisible and tries to explain the transformations of matter. Some time ago at Collioure they used to sell a postcard showing the famous belfry of the little Catalan village that inspired generations of emulators of Derain and Matisse. A holiday painter set up his easel quite seriously before the traditional subject, but the picture that he painted showed the reverse side of the scene, the side of the church that he could not see. Similarly, the chemist describes the hidden side of things: Atoms and molecules can be seen only with the mind.

Louis Pasteur is not only the brilliant explorer of the world of microbes: We cannot forget that he was first a chemist. At the age of 26, in 1848, he made one of the most important discoveries in his century, a discovery whose consequences in numerous fields continue to intrigue us and to inspire research in many directions.

"The molecular dissymmetry that we have established," he writes, "is one of the highest chapters of science, completely unforeseen, opening up totally new horizons to physiology—horizons that are remote, but certain." And since modesty, even false, is better than immodesty, he goes on to say: "I pronounce this judgment on the results of my own studies without any trace of self-satisfaction."

Molecular dissymmetry? The concept is simple, even though the expression may be unappealing: *In the case of certain molecules, in chemistry, their mirror image does not exactly match them.* A manufacturing accident observed by a man who dealt in chemicals resulted in Pasteur's running into one of those molecules that can exist in two twin forms in intimate association in a crystal. He pulled off the trick of separating them. This separation is called a *resolution.* When passed through by a ray of special light vibrating only in a single plane, a solution of one of these forms rotated the plane of vibration to the right, whereas a solution of the other rotated it to the left. Except for this detail, they were, chemically speaking, authentic doubles.

This molecular dissymmetry that Pasteur had detected for the first time in natural organic substances, established, to his mind, "the only clear-cut line of demarcation that one might draw between the chemistry of inanimate nature and the chemistry of living matter." Actually, among the innumerable products that Nature makes, it generally provides us exclusively with one *or* the other of the two forms in which a molecule is conceivable and "realizable." How does Nature do this? So far, it has managed to keep it a secret. The chemist, on the other hand, knows only how to prepare a mixture of the two false twins; then, sometimes with a great deal of effort, the two members of the pair have to be separated: the mixture has to be *resolved*, as chemists say. We shall attempt to show what this operation rests on, and especially why it can be so important, from the point of view of the pharmaceutical industry in particular.

Two "identical" molecules, one of which is right-handed and the other left-handed, are in fact *different*; in particular, they do not have the same biological properties. A dissymmetric molecule, such as a drug or a pesticide, when placed in the presence of one of the molecules that make up a living organism, whether vegetable or animal, which are themselves dissymmetric, does not react with it in the same way that its mirror image does. One does not casually shake either the right or the left hand of a friend: Mothers who belonged to the old school demanded that their children say hello to the gentleman or the lady "with the proper hand."

We shall see that in certain cases, of the two molecules that are mirror images of each other (called *enantiomers*, or optical antipodes), only one has the desired therapeutic activity, while the other is totally devoid of it; yet it can also happen that one has an action that in no way resembles that of the other. Nevertheless, today the majority of resolvable drugs are still offered and prescribed in a *racemic* form, that is, one made up of equal parts of the two enantiomeric species.

Henceforward, pharmacology and the pharmaceutical industry can no longer ignore molecular dissymmetry: the fact that a therapeutically active molecule exists in resolvable forms too often determines its effectiveness, not to mention its side effects. A complete prior study of the curative properties of *both* separate enantiomers will unquestionably be recognized more and more as necessary. One day soon, no doubt, the pharmaceutical industry will have to offer the public only the active molecules, divested of their useless or problematic "doubles." And this means, in fact, that the shadow of Pasteur is already hovering over future drug costs.

These are some of the points that we shall attempt to develop with respect to their theoretical, practical and economic consequences. We shall not, however, lose sight of the fact that the problems in question go well beyond the field of chemistry.

We now know from particle physics that even the electron exhibits a dissymmetry of the same type with respect to the positron—the positively charged particle that constitutes its "match." This is no longer a "curiosity" of interest only to chemists, for here we are dealing with a question involving the fundamental structure of matter in its relationship to what is called "antimatter." Astrophysicists, in their cosmogonic scenarios, are enthralled by this dissymmetry, recognizing it as decisive both for testing out the big bang theory and for checking their hypotheses regarding the origin of life on Earth. It is true, however, that they often confine themselves to approaching this problem only by way of allusion.

One need be neither a great geometrician nor a great observer in order to recognize that the objects that surround us can, from a certain point of view, be classified into two groups: those that are symmetrical and those that are not. The impression of symmetry may be produced by a variety of operations involving the repetition of motifs: Certain tapestries, kaleidoscope images, or a soccer ball made up of polygons in two colors are a few typical examples. A figure is symmetrical when it has, for example, a plane of symmetry that cuts it in two in such a way that every point located on one of the two sides of that plane is exactly matched by a point located on the other side. An asymmetric or dissymmetric object, on the other hand, is such that its mirror image is recognizable, but not exactly identical to it: the two are not superimposable in a point-to-point match.

In addition to the absence of a plane of symmetry, another geometric element can give rise to an object whose mirror image is not superimposable with respect to it. A corkscrew or a boat propeller with three blades has an axis of symmetry: rotation about that axis (by a specific fraction of a revolution that varies according to the case) brings a given point on this object back to a position that makes one forget that there has been any movement.

Properly speaking, the latter objects are not *a-symmetrical* (*a* in the etymological sense of the alpha privative, meaning "not"), inasmuch as they do exhibit a special symmetry and possess an element of symmetry, in this case an axis. This is why rigorous thinkers have replaced the terms "asymmetric" or "dissymmetric" with another, less ambiguous, adjective. The word *chiral*, and the corresponding noun *chirality* (pronounced KYE-ral, kye-RAL-ity) were proposed for the first time by the great English physicist William Thomson, better known as Lord Kelvin. In a lecture dated 1884, but not published until 1906, he said, "I designate as chiral any geometric figure or set of points that is not superimposable on its mirror image." The word comes from the Greek *kheir*, hand (the root of which is found in *chiromancy* or *chiropractor*), which can, as we all know, be either right or left. These new words, following a long period of oblivion, were rediscovered in about 1950 by the subatomic particle physicists, and subsequently adopted at last by chemists. They have given rise to numerous offspring: *achiral* means "not chiral," *homochiral* applies to a set of objects that belong to the *same* right- or left-handed species, etc.

The problems of symmetry, as dealt with by mathematicians, are often far more difficult to understand and explain. However, it should not be necessary for us to develop the preceding remarks any further in order to take up the essential questions that will concern us in this little volume.

The number of molecules present in 18 grams of water is finite and known, but our imagination has trouble grasping its magnitude, unless perhaps we say that it equals the number of teaspoonfuls it would take to fill or empty the Pacific Ocean. Yet that does not prevent chemists from reasoning about atoms and their architectural arrangements as though they were talking about material objects like any others. They do not doubt that the formulas and models they manipulate represent—with varying degrees of perfection, to be sure—a knowable reality. They even sometimes forget, when they write water as H–O–H, that the dash that separates oxygen (O) from a hydrogen (H) actually measures 0.96 angstrom, that is, 0.96×10^{-10}, or 0.96 billionths of a meter.

A collection of these elementary particles, which we can refer to as a pure substance if they are all identical, may occur in any of three physical states: gaseous, liquid, or solid, as the degree of condensation of the matter that is present increases. A crystal represents the solid *par excellence*, with a beauty that attracts one's attention, whereas the sight of steam or fluid stagnating or flowing may leave one indifferent and insensitive. Looking back in time, the very name of Glauber's salt, *sal mirabile* (wonderful salt), discovered in about 1650, might by itself explain the impatience of our curious ancestors to understand that this was really hydrated sodium sulfate. As Gaston Bachelard said, the world must be wondered at before it is understood.

The entities, atoms and molecules which make up matter and whose existence and specific properties chemists have recognized over the past two centuries extend in the three dimensions of space and are thus amenable to geometric description, like crystals, of which they are, one might say, the elementary building blocks.

The study of chiral molecular objects—and of the conditions under which they exist, their names, their manner of preparation and their properties—is one of the liveliest and most important chapters in contemporary chemistry. If we are not too finicky about our vocabulary, we might say that the chemical molecule that interests us cannot be confused with its "double." This word, in fact, which is more suggestive than it is precise, is not the one most commonly used by chemists. They are more given to speak of *enantiomers* (or *enantiomorphs*, from the Greek word *enantios*, opposite) or of *antipodes*. This word, which also comes from Greek (*anti*, against, and *pous, podos*, foot), is an ironic reminder that, as far as chirality is concerned, the human body and its four limbs still serve as a model and a reference.

We can now begin at the beginning, in other words, with the history of the fundamental experiments and initial concepts of a chemistry in three-dimensional space founded by the young chemist Louis Pasteur, before he became better known to the public at large as the legendary discoverer of the rabies vaccine.

I

PASTEUR DISCOVERS "DISSYMMETRY"

The idea that there must be a profound relationship between the form of the ultimate, invisible constituents of a solid and the form in which it appears to us is an intuition that Robert Boyle (1627-1691) already had in a vague way. At the end of the 19th century, Dmitri Mendeleev (1834-1907) expressed this feeling in a more concrete fashion:

"The crystalline form is certainly the expression of the relative disposition of the atoms in molecules and of the molecules in the very mass of the substance. Crystallization is determined by the distribution of the molecules in the direction of their maximum cohesion: thus, the crystalline distribution of matter is itself influenced by the same forces that are at work among molecules; inasmuch as the latter depend on the forces that unite atoms among one another, there must exist an intimate link between atomic composition and the distribution of the atoms in molecules, on the one hand, and the crystalline forms of substances, on the other. Thus, one can judge the composition by the form. Such is the primary idea that constitutes the basis for research on the relationship existing between composition and crystalline forms."

It is not surprising, therefore, that the problems posed by crystalline dissymmetry and molecular dissymmetry were entangled for such a long time. The discovery of "rotatory power" is a magnificent gift given by crystallographers to chemists at the beginning of the 19th century, long before the chemists were able to understand its importance and its consequences. So true is this, in fact, that the discovery of the polarization of light by Étienne-Louis Malus played a capital role in the development of organic chemistry.

ROTATORY POWER

The Danish geometrician and physician Erasmus Bartholin (1625-1698) had observed the "strange refraction of the Iceland crystal" for the first time in 1669. This *Iceland spar*, as it is more commonly called today, which is nothing other than a prettily crystallized form of limestone (what we call calcite, or calcium carbonate), has indeed a curious optical property: When a beam of light passes through a suitably oriented crystal of this transparent mineral, it splits into two rays of equal intensity: a beam referred to as *ordinary* and one known as *extraordinary.*

In 1808, a young French officer who, during Napoleon's Egyptian campaign, had spent his enforced leisure time on problems of optics, "in his house in the rue d'Enfer, used a crystal endowed with double refraction to examine the rays of sunlight reflected by the panes of glass in the Palais du Luxembourg." We will probably never know why Étienne Louis Malus gave himself over to this morning exercise dictated by pure curiosity (but one can deduce from the location of the observer with respect to the Palais, which now houses the Senate, that it had to be on a fine morning). "Instead of the two intense images that he expected to see, he only perceived one, the ordinary image or the extraordinary image, depending on the position occupied by the crystal before his eye. Our friend was quite struck by this strange phenomenon." And for good reason: Malus had just discovered the polarization of light: "Reflection on diaphanous objects, an everyday phenomenon that was as old as the world, had the same property, though no human being had ever suspected it."

"Natural" light vibrates in all directions, perpendicularly to its direction of travel. When this light is turned back by a reflecting surface, it becomes *polarized*: it no longer vibrates in more than a single plane (the "plane of polarization"), which is perpendicular to the reflecting surface. On the day when he made his first observation, Malus verified the generality of the phenomenon at night by examining the light of a candle reflected in a mirror. The astronomer François Arago, who recounted this wonderful story to us, was one of those who followed up on it.

In 1811 (the year before Malus' premature death), in fact, Arago in turn had observed that sheets of rock crystal—pure crystallized silica—cut in a certain way and placed in the path of a beam of polarized light caused the beam to undergo a "remarkable change." When one examines the ray of polarized light that has just passed through the sheet of quartz (another name for rock crystal) by means of a crystal of Iceland spath, one finds that its original plane of vibration is inclined by a certain angle with respect to its initial orientation. A few years later, Jean-Baptiste Biot found that this shift was to the left in the case of some crystals and to the right in the case of others. He also found (in 1815) that liquids of natural origin, such as essence of turpentine, and solutions of sugar or tartaric acid in water, also possessed this "optical activity," this ability to cause the plane of polarization of the light passing through them to rotate in either one direction or the other.

There is a fundamental difference, however, between the origin of this rotatory power and that of certain crystals, such as quartz or potassium chlorate crystals, for example, that Hermann Marback and Eilhard Mitscherlich studied later on. Their optical activity disappears as soon as they are dissolved in water. It was soon obvious (and this was one of Biot's conclusions from the outset) that the action exerted by organic compounds was (in the words of Pasteur himself) a "molecular action, inherent in the ultimate particles, depending on their individual constitution," whereas in quartz "the phenomena results from the mode of aggregation of the crystalline particles." Similarly, the decorative effect of a wall might stem from the ornamentation of each brick or simply from their harmonious arrangement.

The crystal, which combines geometric beauty and dazzling light, is also the preferred meeting ground of two sciences: crystallography and optics. René Just Haüy (who should not be confused with his brother Valentin, who sought, in his way, to restore light to the blind) studied minerals and tried to recognize and classify them on the basis of their multiple symmetries. He observed that rock crystal occurred in two forms that were distinguished only by the existence and the orientation of certain facets. However, the particular distribution of those facets (known by the name *hemihedry or enantiomorphism*) had an important consequence: the two crystalline forms of quartz are *mirror images of one another.*

Bringing together the observations of Biot and those of Haüy, the famous English astronomer John Frederick Herschel (1792-1871), whose illustrious father Friedrich Wilhelm Herschel had discovered the planet Uranus, made another interesting observation: There is a constant relationship between the position of the dissymmetric facets of the quartz crystal and the direction (right or left) in which the polarized light passing through it is rotated.

These, then, were the main events in physics, when he began to engage in his "own work," to use Pasteur's words.

It is almost a tradition in the history of science that the discoveries that open up the finest perspectives lie at the confluence of several disciplines. While the physicists and mineralogists were raising the problems we have just mentioned, the chemists were about to encounter their own.

TARTARIC ACID AND THE MYSTERIOUS RACEMIC ACID

Ever since people learned how to make and drink wine, wine makers have known of the existence of *tartar* in their amphora or casks. The product is already present in grape juice, but is deposited in the barrel only after fermentation: the formation of alcohol diminishes its solubility in water. It was the great Swedish chemist Carl Wilhelm Scheele (1742-1786) who showed, in 1770, that this tartar was a salt resulting from the combination of potassium and tartaric acid; on the same occasion, he described tartaric acid itself. Since that time, tartaric acid has acquired and long maintained considerable importance in industry: In wine-making, it was used for acidifying must so as to improve its fermentation and for the subsequent preservation of the wine. It was employed in the manufacture of chemical yeasts and carbonated beverages. A number of its salts were used in pharmacy, where remarkable properties had long been attributed to them. Seignette's salt, or

Rochelle salt, was a reputed purgative, and the *emetic* (a combination of tartaric acid, potassium and antimony) was recommended as a vomitive as far back as the 17th century. In short, for the budding world of chemistry, for what would later become the field of *pure* chemistry (as opposed to industrial chemistry, which produces large tonnages of mineral acids or plastics, for example), the production of tartaric acid had early on acquired an economic importance that was by no means negligible.

It so happened that in about 1820 one of the little chemical plants at Thann, in Alsace, was the scene of an unforeseen event that was quite remarkable. The owner of the factory, Charles Kestner, witnessed the sudden appearance in his preparations, mixed in with his tartaric acid, of *another* acid whose origin he could not understand and whose strange presence considerably disturbed his usual manufacturing processes. It was first thought that this acid, which was provisionally dubbed *thannic* (from the name of the place where it had originated) was specific to the grapes of Alsace, but this was quickly found not to be the case. The appearance of this importunate acid repeated itself several years in a row, then everything fell back into place and the chemists who had begun to take an interest in the mysterious product subsequently had considerable difficulty renewing their stock. In any event, by mid-century this unseemly substance was the butt of many facetious remarks in the little world of the chemists.

The celebrated chemist and physicist Joseph-Louis Gay-Lussac (1778-1850) was one of the first to study this acid, which he rebaptized *racemic* (*racemus*, from which the word *raisin* is derived, in Latin means a bunch of grapes), an adjective that was to have a brilliant future. Gay-Lussac made a number of important observations concerning it: he noted that the salts (calcium salts, in particular) of this racemic acid were far less soluble than those of tartaric acid itself, but that, as far as the composition of the two acids was concerned, they could not be distinguished on the basis of an analysis of their elements. Crystallographers had managed to observe the appreciable characteristic differences in appearance and form between the two series of acids and their salts.

The observations of Biot in 1838 bring us back to our subject with this major finding: Whereas tartaric acid and its salts acted on polarized light, *racemic acid possessed no "optical activity."*

Finally, apart from this very special difference in their properties with respect to polarized light, the case of tartaric and racemic acids could not really surprise the chemists of the time. It was not without good reason that the Swedish chemist Jöns Jacob Berzelius (1779-1848), one of the oracles of the world of chemistry of his day, had invented the word *isomerism* (from the Greek *isomeres*, composed of equal parts). He had been led to this in connection with the case presented by *cyanic* acid and *fulminic* acid, whose chemical behavior differed considerably despite the fact that they were made up of carbon, hydrogen, oxygen and nitrogen in the same proportions (HCNO and CNOH). Michael Faraday (1791-1867), too, had reported the same isomeric relationships among several hydrocarbons.

At this stage of our story, the curious relationship between tartaric acid and racemic acid was intriguing, to be sure, but did not yet preoccupy more than a few chemists; and while it was not yet understood, the phenomenon of which they offered a new example had already been given a name, for it sometimes happens that scientists console themselves in this way.

MITSCHERLICH'S FRUITFUL ERROR

This enigma could not remain one for long, however. It received a new, fortunate and unexpected impetus, thanks, one might say, to sheer luck and to the mistake of a prominent scientist. It was unforeseen due to the fact that, as we shall see further on, it originated in the observation of a relatively unlikely phenomenon; and it was fruitful because it made the mystery even deeper, so that it became unbearable.

Eilhard Mitscherlich had been an enthusiastic student of the history of the Ghurids and Chorazmians. Later, after he had abandoned those Central Asian peoples and discovered an interest in chemistry, he observed that certain salts exhibited remarkably similar appearances, despite the fact that their chemical compositions were different. This is the case of phosphates and arsenates, or of the sulfates of iron, zinc or magnesium. This analogy of crystalline form, dubbed *isomorphism*, a term derived from Greek, reflects the close relationship among elements constituting several species. This property is manifested in particular by the readily demonstrable fact that two atoms or two groups of atoms can be substituted for one another in the same crystal in any proportions. The discovery of isomorphism played a capital role in the history of modern atomism: It helped suggest the parallels later crowned by Mendeleev's classification.

Within the framework of these investigations, Mitscherlich, in 1844, had studied the double tartrate and racemate of sodium and ammonium. By *double* one must understand, in current terminology, that the two alkaline ions are combined with the same acid molecule. He had ascertained, once again, that these salts did indeed have the same chemical composition, but—and this is what was new—from the crystallographic point of view, this tartrate and this racemate were, to him, *totally indistinguishable*. To this German scientist, in both these salts of sodium and ammonium "the nature, the number and the arrangement of the atoms and the distances between them are the same." Up to that time salts belonging to the two series, tartaric and racemic, had always exhibited forms that were ever so slightly different. Here, then, for the first time, were crystallized isomers that nothing permitted one to distinguish, other than the fact that the tartrate, when dissolved in water, caused polarized light to rotate, which this exceptional racemate did not. Mitscherlich, as we suspect, had not seen right and had not seen everything. Pasteur took it upon himself to find out where the mistake was, and found it.

ENTER PASTEUR

In 1848 Pasteur was 26 years old. The young graduate of the *École Normale Supérieure* had solid training as a chemist and crystallographer, thanks to, among others, Auguste Laurent (1807-1853), one of the pioneers who, not without some difficulty, helped to impose modern atomic theory. Free to choose his own research topics as soon as he had acquired his doctorate in science, he turned his attention to tartaric acid, its salts and the problems they posed—enough to preoccupy a truly curious mind. He very quickly noticed something that had escaped his predecessors. All the crystallized tartrates that he studied "showed unquestionable signs of hemihedral surfaces," those dissymmetric facets discovered by Haüy on rock crystal and recognized by Herschel as related to the existence of rotatory power. The racemates that had no effect on polarized light, on the other hand, never exhibited such hemihedrism, with one exception: the racemate of sodium and ammonium that Mitscherlich had stumbled upon.

We can guess the rest: Pasteur, on examining the wayward salt more closely, found that it was in fact made up of a mixture of crystals of two types in equal quantities, some exhibiting left-handed hemihedrism, and others, right-handed. A subtle distinction indeed, and one can understand that, in order to make it, it took a good pair of eyes and the intense desire to perceive it. The two nonsuperimposable crystalline forms were mirror images of one another: the first belonged to a "natural" tartaric acid salt known since Scheele, and the second, whose existence had been unsuspected until then, was Pasteur's discovery. With tweezers he picked apart the two species under the initially skeptical but quickly enthusiastic gaze of Biot himself, who was brought in to witness this fundamental discovery. A solution of one of the species in water did indeed "rotate" the plane of polarized light to the left, while a solution of the other turned it in the opposite direction.

One cannot but wonder at the combination of good luck and sagacity that enabled Pasteur to observe this *spontaneous resolution* of the racemate that had escaped the notice of Mitscherlich.

Indeed, it in no way detracts from Pasteur's genius if we remind ourselves of his first stroke of luck. Now that we are familiar with thousands of crystalline "racemic" compounds, namely, compounds such as those in Pasteur's historic example, consisting in equal parts of two molecular species—a right-handed and a left-handed one—we know that *in nine cases out of ten such spontaneous resolution cannot occur.* A hundred years later, in the 1940s, only twenty or so of these special cases were known to exist. And even today, though we know of several hundred new separations of this kind, most of them were not detected with "Pasteurian" eyes, if we may say so. Indeed, the existence of the hemihedrism that alerted the great man is not something widespread at all, and even when it does exist, it often remains, for practical reasons, very difficult to demonstrate.

Pasteur shared his second stroke of luck with Mitscherlich: They managed to obtain crystals of double sodium ammonium tartrate that measured several millimeters, and in fact several centimeters. Not all chemical substances enjoy such growth and crystallize so nicely; some crystals, despite every sort of technical artifice, remain stubbornly microscopic. Under such conditions, in other words, it would have been out of the question to recognize the famous little dissymmetric facets with the naked eye and manually sort the two forms present.

Pasteur had the advantage of still a third favorable circumstance. Much later, the great physical chemist Jacobus Hendricus van't Hoff (of whom we shall have occasion to speak in glowing terms further on) showed that the first historical case of spontaneous resolution *was observable only under very special temperature conditions.* A few degrees up or down, and Pasteur would not have been able to see anything. The fact is that above 28°C the mixture of equal parts of clearly distinct, mechanically separable right and left hemihedral crystals gives way to *another crystalline variety* in which the two species of substances are, one might say, combined, complementarily associated in the same solid lattice, as is the case with racemic acid itself. In short, if instead of crystallizing his sodium ammonium racemate in a Paris basement he had worked in a laboratory located in Dakar, Pasteur would no doubt never have become the Pasteur who founded *stereochemistry*, the science of molecules in their three-dimensional arrangement.

A SECOND STROKE OF LUCK AND GENIUS

Later, after he had become professor at Lille, he would tell us, in an official speech, in connection with the fortuitous discovery of electromagnetism by the Danish physicist and chemist Hans Christian Ørsted, that *chance favors the prepared mind.* How could Pasteur, in making this famous statement, not be thinking first and foremost of his own scientific adventures?

It had in fact not taken him very long to observe, when examining the crystalline forms of right-handed and left-handed tartaric acid salts prepared with other metallic ions, that they did not all differ in the position of their dissymmetric facets. There are many cases, he had noted, in which "*perfect and absolute identity* exists between the crystalline forms" of the two inverse series, which thus become totally indistinguishable. What could be done to get around this obstacle to the generalization of his initial resolution, performed so brilliantly?

"I think that in cases where the inherent crystalline structure of substances active with respect to polarized light was not visibly and geometrically evident, it would suffice to alter the conditions of crystallization in order to cause the hemihedral facets necessarily and consistently to appear."

In the numerous tests that this conviction suggested to him, Pasteur went on to study the crystals of salts formed by his right-handed and left-handed tartaric acids with natural "organic alkalis," such as *cinchonine*—an alkaloid, as we would call it today, that is a close relative of *quinine* and is also extracted from cinchona bark. As expected, he immediately recognized the "character of nonsuperimposable hemihedry." But that was not all, and what he next discovered went beyond anything he might have hoped for: these observations opened up for him the prospect of a new method of separating the two acids that coexist in racemic acid. The two tartrates (right and left) of cinchonine differ not only in terms of their crystallographic aspects, but also their *solubilities*. In particular, the right-handed tartrate of cinchonine very readily dissolves in alcohol, whereas the left-handed tartrate is highly insoluble in it. What is more, an analysis of the right tartrate of cinchonine reveals that it contains *eight* water molecules, whereas the left one contains only *two*.

In short, the right-handed tartrate of cinchonine *is not* the image of the left-handed tartrate of cinchonine, and Pasteur tells us why:

"Let us combine a dissymmetric substance with a substance having a plane of symmetry. Let us suppose, for example, that with my right hand I hold this book. This results in an assemblage that is thoroughly similar and not superimposable on the assemblage that we would obtain if I held this same book in the same manner with my left hand.

"But let us imagine the assemblage of a dissymmetric substance with a dissymmetric substance; let us suppose, for example, that I take a human foot in my right hand. This assemblage will no longer have simple, but double, dissymmetry, which will be quite different, in the last analysis, from the dissymmetry of the assemblage of my left hand with the same foot. In one case, the right-handed dissymmetry of my right hand will be added to the right-handed dissymmetry of the foot, if it is the right foot, whereas when my left hand is combined with that right foot, the two dissymmetries will work against one another. And inasmuch as there is a right hand and a left hand, a right foot and a left foot, four assemblages will be possible: right hand, right foot; left hand, left foot; right hand, left foot; left hand, right foot."

To continue the Pasteurian metaphor, if natural cinchonine is symbolically represented by a right hand, it selects its partner without hesitating, grasping the right foot *or* the left foot, the two enantiomers present in racemic acid.

In addition to cinchonine tartrates, Pasteur studied salts formed from other natural "bases," which had long been well known to pharmacists and apothecaries, such as *quinine*, *brucine* and *strychnine*. These variations led him, in his own terms, to the "same general results" as with cinchonine itself.

It was not long before these observations led to further fruitful implications. The first was self-evident (or nearly so): by treating racemic acid with a suitable alkaloid, he was able by selective crystallization to obtain, in the pure state, the less soluble of the two simultaneously formed salts. This was a new method of resolution that made it possible to obtain the right and left tartaric acids in a manner far simpler and more general than the manual sorting he had performed on the sodium ammonium racemate. The findings obtained with cinchonine salts, however, were to lead him to other results of primary importance, albeit in a less direct way, providing us with a new example of the tortuous ways that science sometimes follows in leading to discovery.

THE DISCOVERY OF RACEMIZATION

It is truly exciting for a researcher or for a philosopher of contemporary science to try to reconstruct the itinerary followed by a scientist who, going along in the dark, finds his or her way to something absolutely novel (so to speak), to which no hypothesis, no theory could have served as a guide. It is a stimulating exercise even if, in the absence of any precise document, such reconstruction involves some risks. How and why did Pasteur discover that, while he had been able to go from racemic acid to its two component tartaric acids, another road could lead him from tartaric acid to racemic acid?

"The right-handed tartrate (of cinchonine) loses its water and starts to take on color at 100°C; the left-handed tartrate also loses its water of crystallization at 100°C and from that point on it is perfectly isomeric with the right-handed tartrate, yet can withstand a temperature of 140°C without becoming colored." The rest of the experiment proves even more astonishing: "As this salt is subjected to gradually increasing temperatures," Pasteur first finds that the alkaloid becomes altered (to the point of actually changing structure, but that is another story). Soon, however, "... tartaric acid, too, undergoes considerable changes; ... *after five or six hours at a sustained temperature of 170°C, ... The flask is broken.* The *black, resinous mass contained in it* is treated several times with boiling water, and to the filtered liquid, after cooling, an excess of calcium chloride is added, which immediately precipitates all the racemic acid in the state of lime racemate, from which the racemic acid can easily be extracted."

A few years later, Victor Dessaignes (1800-1885) observed that he could obtain racemic acid from tartaric acid under conditions not quite so violent. It sufficed to heat the solution for a fairly long time in dilute hydrochloric acid or even in plain water. These were the very circumstances under which thannic acid had mysteriously appeared in the vats of the Kestner plant some thirty years earlier. At last it became clear that it was the overheating of industrial tartar that had caused the incident from which science was to derive inestimable benefits.

Pasteur, at first glance, had no good reason to heat the cinchonine tartrates whose right-handed form, he had just observed, contained eight molecules of water, while the left-handed form contained only two. No obvious reason, at least, to heat them to the point of decomposing them for the most part, since it was enough for him to ascertain that this "water of crystallization" was indeed part of the intimate organization of the crystals and was there, quite simply, only due to their imperfect drying. Actually, as we shall see, in undertaking this barbaric but ultimately effective operation, he had an ulterior motive.

Be that as it may, we know today that this discovery of the *racemization* of tartaric acid has broad implications: under certain structural conditions, a considerable number of chiral molecules are *racemizable*, or in other words, one of their *enantiomers* (the name given to the pure right- *or* left-handed form) can be transformed into a *racemic* mixture of the two.

A MISTAKE OF PASTEUR'S

During the violent treatment of cinchonine tartrates, Pasteur had not obtained only racemic acid. It was accompanied by *another* acid that was an isomer of all those he had found so far. This new acid, whose crystallographic properties (as well as the crystallographic properties of its salts) were specific, had no effect on polarized light, but, unlike racemic acid, *it could not be resolved.* Pasteur called it *inactive tartaric acid.*

This new finding led Pasteur, who was less fortunate in this episode than in other chapters of his scientific adventure, to erroneous conclusions: The history of science is also written with the errors of geniuses, which sometimes contain interesting lessons. To understand this wrong move, however, we must go back briefly in time.

Aspartic acid is easily prepared from *asparagine*, a natural product that can be extracted from young asparagus (as its name indicates); *malic* acid comes from the apple (called *malum* in Latin). Pasteur had shown that these two substances were active on polarized light. Also, thanks to the Italian Rafaele Piria (1815-1865) and to Victor Dessaignes, the "clever chemist of Vendôme," of whom we have already spoken, it was possible to transform aspartic acid into malic acid and vice versa (by simple reactions that we need not go into). Now, starting with substances characterized by rotatory power, Pasteur found that the products obtained at the end of these transformations were optically inactive (did not cause the plane of polarized light to rotate). Pasteur, obviously, had not failed to ask himself a question which he felt it was "easy to answer": "Are not these inactive aspartic and malic acids combinations of right- and left-handed acids analogous to racemic acid?" He did not hesitate to answer *no*, "That view cannot be upheld." Let us skip over the experimental facts that he erroneously put forward in support of his views—history would subsequently show that they had been incorrectly observed—but let us retain the theory on which he based them.

"It is rational to think that a dissymmetrically constituted molecular arrangement that is *subjected to the action of a high temperature* might change into another molecular arrangement in which the special disposition that produced the dissymmetry of the original arrangement has disappeared." (The detail which I have highlighted of course explains the severe treatment to which he had subjected his cinchonine tartrate, for which we were seeking to understand the reason.)

Inactive malic acid and aspartic acid were thus, to his mind, *untwisted*, to use his own expression. "The natural acid," he explains, "is a winding stairway for the arrangement of the atoms; [the inactive acid] is the same stairway consisting of the same stairs, but straight instead of in a spiral."

Again he says: "While I had been quite unwilling to seek the transformation of tartaric acid into racemic acid, I performed numerous tests to arrive at inactive tartaric acid. Not only did it seem to me to have a possible existence, in view of the theoretical ideas; I also knew the whole close link between tartaric and malic acids and had previously obtained inactive malic acid."

Thus, we end this chapter with a question that the theory of asymmetric carbon will shortly answer: Why was Pasteur mistaken when he affirmed the existence of a form of malic and aspartic acids that was "naturally inactive," whereas such a form does indeed exist in the case of tartaric acid?

II

ASYMMETRIC CARBON AND MOLECULAR CHIRALITY

ATOMS AND MOLECULES IN ABOUT 1850

If the reader has not found, in our exposition of the work of Pasteur and his contemporaries, any formulas that would have enabled any high-school senior better to understand certain points, it is not only out of compliance with the spirit of this series, which seeks to minimize the use of dull scientific formulas (whether mathematical or chemical). The reason is simpler: In the rare reactions written out by Pasteur (for example, in his note of 1852 on malic and aspartic acids), water is formulated as HO, and not H_2O. In other words, "his" chemistry is often more difficult for us to "read" than his prose, which is always clear within the limits of the science of his time.

Real light will be shed on Pasteur's observations and their consequences only by the *theory of structure in chemistry*, the foundations of which were laid by Archibald Scott Couper (1831-1892), August Kekulé von Stradonitz (1829-1896) and Aleksandr Mikhaylovich Butlerov (1828-1886) starting in 1857-

1858, and especially by *the theory of asymmetric carbon*, a few years later.

These theories, it must be remembered, were not accepted at first, any more than was the possibility of the existence of atoms. The battle—and indeed it was a battle—between *atomists* and *equivalentists*, between those who believed in atoms and those who did not, occupied French chemistry for nearly forty years. Illustrious and influential scientists like Marcelin Berthelot (1827-1907) or Henri-Étienne Sainte-Claire Deville (1818-1881) imposed their skepticism and their blindness on their students, who, later on, in turn became teachers. It can be said, without too much exaggeration, that those harsh battles left traces that it took several decades to heal, especially in education and in the training of researchers.

Before turning to the solutions that atomic theory provided for the problems raised by the young Pasteur, it would perhaps be worth our while to situate these controversies more clearly in the history of ideas and in their philosophical perspectives. To be sure, the protagonists of these polemics rarely, as far as I know, gave a clear statement of the ideological background that lay behind their contradictory views. I feel, however, that based on the new data that they accumulated, the chemists of the past century might have claimed kinship with either of two major intellectual families.

For those who did not believe in atoms, the reality that they manipulated remained hidden and inaccessible; it was truly Kant's thing-in-itself. Only relationships between phenomena were knowable. For these scholars of science, the chemical formulas which they used were not pure symbols, a system of representation that sought to summarize the greatest possible number of facts in a coherent way. Their ideal science closely resembled that defined by the philosopher Étienne Bonnot de Condillac (1714-1780): a well-made language. Mistrust of theories almost naturally constitutes a component of this attitude.

The attitude of the other group of chemists was the opposite of this tendency. They were convinced that it was possible to know the material reality that they observed and transformed. To them, atoms and molecules were genuine objects, and they wanted, sometimes with great naïveté, to undertake the description and study of the mechanisms governing their transformation. Chemical formulas themselves were not only abstract, conventional signs; they viewed them as veritable diagrams or plans, the same way that architects or engineers view the blueprints on which they work. This ambition bore the marks of an imagination that was sometimes given too free a reign and was not far off from a certain dated romanticism: It obviously involved the risk of going off the track, which some of them did not manage to escape doing.

To anyone who looks back on this debate with the perspective afforded by a century and a half, it is obvious that the philosophy of Auguste Comte (1798-1857) runs through the entire fabric of this affair: positivism reduced to its bare essentials, recommending first of all that one refrain from any "vain speculation about causes," a simplified philosophy of science that seeks to stick to realities "perceptible to our organism" (namely, to phenomena perceived by our senses) and the laws that govern them. But a "minimum" positivism against what adversary? An enemy perhaps just as difficult to pinpoint, inasmuch as modern atomism during its difficult adolescence did not claim adherence to, or seek to hide behind, any specific philosophy, in line, for example, with Greek materialism, which Lucretius (ca. 98-ca. 55 B.C.) had popularized. The "positive" science that Jean-Baptiste-André Dumas (1800-1884) or Marcelin Berthelot held up against those who believed in atoms and the possibility of knowing how they were arranged was thus levelled more against a state of mind than an actual doctrine. Actually, outside of a clearly defined philosophical and ideological framework, this episode can be reduced to a still more rudimentary scheme: open science as opposed to closed science, science open toward the future and adventure as against science closed in its habits and its taboos.

There can be no question here of going back over the detailed history of the discussions that occupied front and center on the stage of chemistry in the 1850s, at the time when Pasteur slipped in by the back door reserved for crystallographers. We might, however, summarize them as follows (if we do not shrink from such an oversimplification):

If atoms are knowable, what are the precise weight relationships that exist among the different chemical elements that make up matter? How does one determine atomic masses, the same ones that we find today by simply consulting Mendeleev's well-known table?

This preliminary question was not answered until the end of the 1850s (and even then, not for everyone). It then remained to ascertain what rules governed the concatenations of atoms in their various combinations (in an organic molecule, in particular) in order finally to constitute a *structure* that was also accessible to the consciousness.

The final question leads us right to the heart of our subject: Can such rules of concatenation enable us to understand the existence of isomeric molecules, some having and others devoid of rotatory power?

LE BEL, VAN'T HOFF AND ASYMMETRIC CARBON

The notion that organic compounds are the result of a concatenation of carbon atoms having four combining capacities, four *valences*, each of which can be satisfied by an element such as hydrogen or chlorine, for example, goes back to the years 1857-1858.

This theory, formulated almost simultaneously by a German, August Kekulé, and the Scotsman Archibald Scott Couper, lies at the basis of the notion of chemical *structure* developed by the Russian chemist Aleksandr Butlerov. It enables one to understand, in particular, how, starting with a given number of atoms linked in different ways, it is possible to obtain several different *isomers*, the number of which can be predicted. Thus, to take only a simple example, starting with three carbon atoms, eight hydrogen atoms and one oxygen atom, one can construct the following two arrangements:

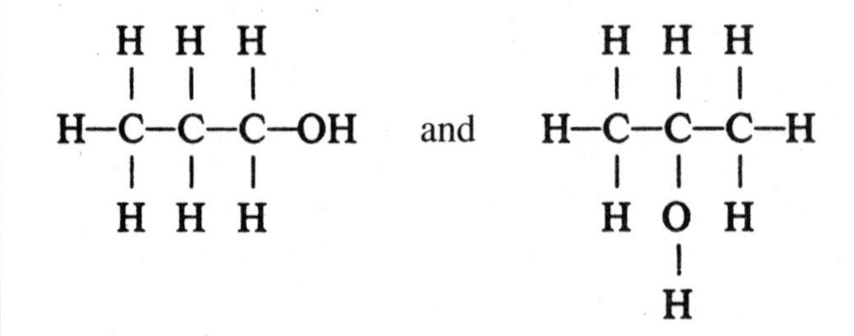

But let us come to the point which constitutes a considerable advance with respect to the subject that concerns us.

Let us imagine that these "affinities" of the carbon atom are directed toward the apices of a tetrahedron in which it occupies the center. And let us then imagine that these four bonding possibilities are satisfied by four atoms of different types: one chlorine (Cl), one iodine (I), one fluorine (F) and one hydrogen (H), for example. We thus have a molecule *that can exist in two nonsuperimposable forms, mirror images of one another*, which is not the case when a molecule contains at least two identical "substituents."

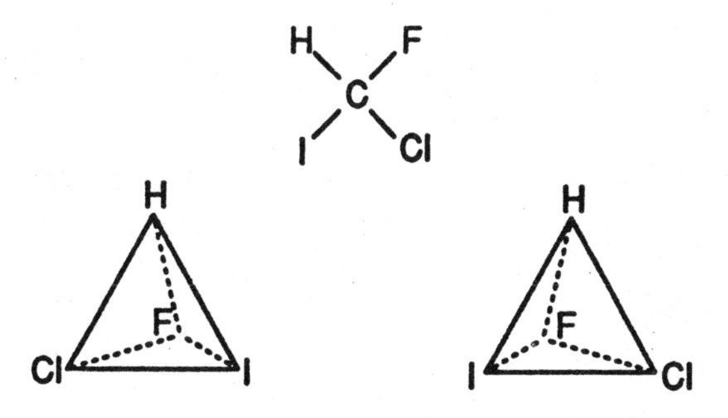

Such, in its most elementary form, are the essentials of the *asymmetric carbon theory*, formulated independently in 1874 by the Frenchman Joseph-Achille Le Bel and the Dutchman Jacobus Hendricus van't Hoff, two young chemists who had become acquainted a few years earlier in Paris, in the laboratory of Adolphe Wurtz, part of the inner circle of the avant-garde science of the time. They had not, however, shared their common preoccupations with one another.

Le Bel claimed in private, but without much boasting, to be descended from a Le Bel who had waited on Louis XV, at the Parc-aux-Cerfs. His family owned the oil shale deposits at Pechelbronn, in Alsace. An Engineer of the École Polytechnique, after being trained in chemistry with the best teachers he devoted himself, starting in 1874, to the exploitation and development of the family's petroleum. But he had already found the time to publish the paper that made him famous, *On the relations that exist between the formulas of organic compounds and the rotatory power of their solutions*. Being free of all material concerns, he managed throughout his life to affirm his freedom of thought, vis-à-vis both the government authorities and recognized professional classifications. This wealthy nonconformist bachelor, who had his own private laboratory in the Latin Quarter, was interested in chemistry as well as geology and cosmology. He presented his candidacy to the Academy of Science unsuccessfully more than once, then finally managed, in 1929, one year before his death, to occupy the vacancy left by Marshal Foch.

The career of his friend van't Hoff similarly followed nonclassic lines. He was the son of a Rotterdam physician, and had chosen to learn chemistry, first in Bonn and then in Paris. He returned to the Netherlands in 1874, where he published, in Dutch, his treatise *On Structural Space Formulas*, which made him, together with Le Bel, one of the founders of stereochemistry. After a university career that began at the Veterinary School of Utrecht, he died of consumption in 1911, at which time he was a professor in Berlin. Even to those who admired him, van't Hoff was a poor mathematician, an experimenter of no great talent and an uninspiring professor. Yet that did not prevent him from laying the definitive foundations of certain chapters of physical chemistry "with instruments that a 'serious' physicist would not have accepted even for preliminary experiments."

The French version of his historical tract on space formulas (or *perspective formulas*), which was translated immediately, has not developed a wrinkle with time and remains a model of simplicity and clarity.

"In the case where the affinities of a carbon atom are saturated by four groups which differ among themselves, one can obtain two different tetrahedrons that are mirror images of each other and can never be superimposable; in other words, we are dealing with two isomers in space."

Le Bel's fundamental paper, which led to the same ingenious explanatory theory, was the product of an entirely different mind. It was presented with a thoroughly mathematical rigor as the statement of a process of "reasoning that enables one to arrive at a law." A comparison of these two texts would be an exciting exercise for professors of philosophy who needed to illustrate their lectures on the *geometric mind* and the *subtle mind* and on "style" in the field of scientific discovery.

Whatever the case may be, the asymmetric carbon theory provides an immediate, clear answer to several of the questions raised by Pasteur's results.

To start with, on the question raised by the existence of *untwisted* malic and aspartic acids, van't Hoff replies: "Let us consider the structural formulas of these two compounds: they have only *a single* asymmetric carbon. Thus, for each of these substances, one can predict only *two* isomers, one right- and one left-handed. Their so-called *untwisted* form does not exist, and was confused with a racemic mixture of the two entities, right and left." Indeed, it will be demonstrated later that these untwisted acids can be obtained by mixing the two corresponding enantiomers, which, conversely, may result from the resolution of these new entities which existed only in Pasteur's imagination.

Untwisted tartaric acid, which, on the other hand, is naturally inactive, cannot be confused with racemic tartaric acid: here Pasteur was right. Le Bel and van't Hoff can now explain to us why. We know that the tartaric acid molecule contains two neighboring carbons that are asymmetrical since, in addition to the bond that unites them, each is surrounded by two different groups (which, for the sake of simplicity, we shall designate as A, B, and H). The two enantiomeric tartaric acids can thus be represented by the following two figures:

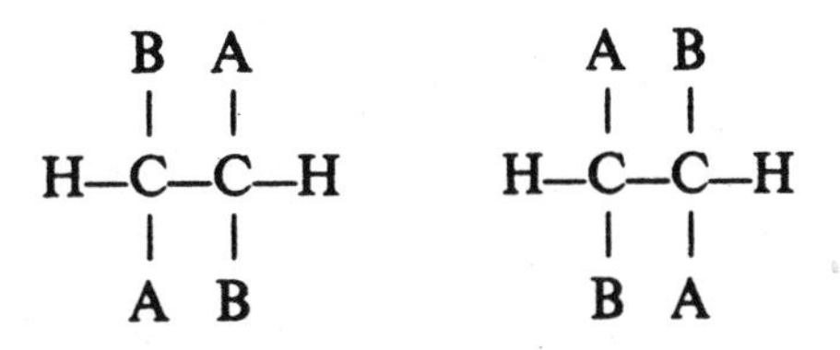

A different arrangement of the groups surrounding the two central atoms, on the other hand, gives rise to a different molecule, which is symmetrical and consequently superimposable with respect to its mirror image:

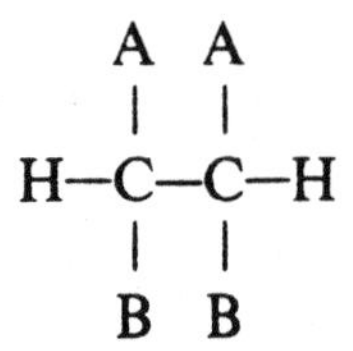

This, therefore, is indeed Pasteur's unresolvable inactive acid.

The explanation of the isomeric possibilities offered by tartaric acid with its two asymmetric carbons obviously has general validity.

Van't Hoff, in the pamphlet-manifesto that he published in 1887, was already considering the far more complex case of the various natural sugars and their structures. The most common ones belonged to the family of *hexoses*, of which *glucose* is one of the best-known members; they have six carbon atoms (whence their name: *hex* is six in Greek), four of which are asymmetric. These "carbohydrates" (hydrates of carbon) combine the elements of water around carbons linked in a chain in the center of their molecules. Simplifying to the greatest possible extent, we can attribute to them the following general formula (without further specifying the nature of the terminal carbon atoms A and B):

```
  H  OH  H  OH  H  OH  H  OH
   \/     \/     \/     \/
A—C——————C——————C——————C—B
```

If in a purely arbitrary way we attribute the sign + or – to each of the asymmetric carbons present in these molecules, we obtain 16 different combinations: eight of them are images of the other eight. Van't Hoff pointed out that the number of possible isomers could be computed from the number (n) of asymmetric carbons present; it is given by the formula 2^n.

In the event that the two terminal carbons (which we have designated as A and B) are the same (A = B), some of the preceding possibilities disappear: certain combinations acquire a plane of symmetry that passes between the two central carbon atoms, thus losing their chirality. In this particular situation, "sugars" of this type are "by nature" deprived of rotatory power, just as in the case of inactive tartaric acid, which we examined earlier. The question still remained as to which of these 16 predicted structures corresponded to the eight hexoses found in nature; which one, for example, belongs to natural glucose, which has a positive rotatory power? This problem was not solved for another seventy years, when the notion of absolute configuration was finally elucidated.

In any event, to come back to our tartaric acid, Pasteur remained strangely silent regarding the asymmetric carbon theory, which provided such clear answers to the questions he had raised some twenty years earlier. He did not say a word about it in a famous lecture given before the Chemical Society of France in 1883—in other words, nearly ten years after the appearance of the illuminating work of Le Bel and van't Hoff in the field he had discovered and so masterfully pioneered. To be sure, he had the excuse that in the meantime he had turned his attention toward new concerns. But if he was wrong to ignore asymmetric carbon, at least he said nothing against it. As we shall see, other, less judicious, individuals did not show the same wisdom.

THOSE WHO DID NOT BELIEVE IN ATOMS

The "chemistry in space" christened by van't Hoff did not arouse widespread enthusiasm—far from it. Now, maliciousness is not the only reason that makes it interesting to dig up the violent, pretentious criticisms levelled against it: they sometimes came from well-established scientists, not all of whom were third-rate. We mention them because they illustrate, almost to the point of caricature, the excessive self-confidence indulged in by certain individuals who, sheltered behind *their* science, talk about things they are ignorant of or fail to understand. We will also find here the traces of that positivism referred to above, though it is rarely owned up to.

As of the first paper on the asymmetric carbon theory, presented before the Chemical Society of France in 1874, Marcelin Berthelot felt compelled to react, speaking down from his perch of authority:

"Mr. Berthelot presents various observations on the structural formulas in space proposed by Mr. van't Hoff. While we do not fail to recognize, generally speaking, the advantages of such formulas, which are more rational than the flat formulas ordinarily used, it must nevertheless be noted that there can be no complete representation of chemical compounds without bringing in the notion of the rotatory movements and vibrations that animate each atom in particular and each group of atoms in the molecule.

"Now, this dynamic notion of compounds leads, in a large number of cases, to simpler, more general explanations than the purely geometric representation of atoms laid on a flat surface or even distributed among the apices of a polyhedron.

"Thus, the existence of four molecular types: the right-handed type, the left-handed type, the neutral type (combination of the two preceding ones), and the inactive type, which seem to belong to any substance endowed with rotatory power, does not appear to be adequately explained by purely static considerations such as those just presented. Indeed, there can be no doubt that these four types belong to a single atomic structure, or at most to two symmetrical structures, such as those proposed by Mr. Le Bel and Mr. van't Hoff. This is true whenever the four types have the same manner of synthetic formation and the same analytical reactions.

"The existence of rotatory power is clearly explained by the different orientation of the molecules; or, more precisely, by the orientation of their vibratory movements. Within a given molecular system, one can indeed conceive that certain of its atoms may all vibrate in the same plane, which gives us the inactive substance; or else in another plane, inclined symmetrically to the right or the left with respect to that of the fundamental atoms, thus yielding the right- or left-handed substance. One can also conceive of two symmetrical systems juxtaposed in such a way as to yield an intermediate state of movement analogous to that of the inactive compound, which is the case of the neutral substance. The four types of substances characterized by rotatory power are thus explained in this way. Also conceivable is the existence of a multitude of isomeric substances with atomic structures that are similar but differ in the unequal and dissymmetric orientation of the vibratory movements of their atoms."

Resuming the offensive in 1876, Berthelot reproached the theory of Le Bel and van't Hoff by predicting that a certain hydrocarbon contained in essence of *styrax* or *storax* (an aromatic plant used in pharmacy), to which the name *styrolene* had been given, *could not* possess rotatory power due to the fact that it had no asymmetric carbon. He, Berthelot, however, had prepared a sample of it in 1867 that deflected polarized light. He went on to devote several accounts to this surprising property of his styrolene, concluding that any theory incompatible with the facts he believed he had demonstrated was "by that very fact proven incorrect. Negations and limits are the touchstones of theories." Need we say that it was Berthelot's results that contained the errors? Actually, the rotatory power that he had observed and had attributed to styrolene was due to an impurity contained in natural essence of styrax—an impurity which, in spite of his repeated distillations, he had not managed to eliminate.

In another area, a few years later, in 1877, Hermann Kolbe (1818-1884), a German chemist of no mean repute, continued the battle in his *Journal für praktische Chemie*, of which he was the editor-in-chief. Needless to say, his pamphlet is surprising more for its sharp tone than for the richness of his argumentation:

"A certain Dr. J.H. van't Hoff, employed at the Veterinary School of Utrecht, had, it would seem, no inclination for exact chemical research. He deemed it more convenient to climb on the back of Pegasus (borrowed, obviously, from the veterinary school) and proclaim in his *Chemistry in Space* that after a bold flight there had appeared to him, from the top of Mount Parnassus, atoms arranged in space.

"As the prosaic world of the chemist has little taste for this type of hallucination, Dr. F. Hermann, an assistant at the Heidelberg Institute of Agriculture, undertook to give it broader dissemination with a German edition. ...

"It is indeed a sign of the times, which suffer from a lack of critical spirit and an aversion to criticism, that two unknown chemists, one from a veterinary school and the other from an institute of agriculture, can decide with such assurance the most difficult problems of chemistry, which no one will be able to resolve—in particular this problem of the arrangement of atoms in space—and present their solutions with a nerve that leaves the real scientists flabbergasted."

Finally, we must say a word about another "refractory case," that of Gregory Wyrouboff (1843-1913), a personage long since forgotten whose positions, which were more subtle and more difficult to grasp, were in conflict both with Pasteur's ideas and with those of Le Bel and van't Hoff, since they involved the relationship between molecular asymmetry and crystalline asymmetry.

Wyrouboff was a crystallographer of Russian origin who had taken up residence in Paris. A convert to militant positivism, he had founded— together with the father of the celebrated dictionary, Émile Littré (1801-1881)—the *Revue de philosophie positive*, of which he was the manager from 1867 to 1881. On the death of his friend, and "unlike many scientists, who start with invention and end with philosophy or politics, he returned wholeheartedly to the scientific studies that he had neglected." But he was to be unfaithful to them again in 1903: at that time he was appointed professor of the History of Science at the Collège de France, having been ultimately preferred by the government decision-makers of the time over his unquestionable competitor, Paul Tannery (1848-1904).

Let us endeavor to follow the arguments that he developed in 1894 before the Chemical Society of France in order to contest the asymmetric carbon theory that had twenty years of success behind it at the time:

"Mr. Wyrouboff shows, in connection with the recent theory of Mr. Freundler, that the conception of chemists regarding symmetry diverges considerably from that established by geometry. ...

"One can easily demonstrate that no polyhedron can be constructed with dissimilar elements; and that this being the case, different atoms or groups of atoms cannot enter into the structure of a polyhedric molecule possessing symmetry. Whatever may be the difference that one establishes between crystals and chemical molecules, it will always be necessary, ultimately, to construct the former with the latter; naturally, the symmetry of the former will therefore have to be identical with the symmetry of the latter. It is highly regrettable that the chemists, when imagining the tetrahedron theory, did not think of this fundamental condition of molecular physics."

To this, the excellent chemist August Béhal (1859-1941) hastened to reply:

"There is no simple relationship between the form of the chemical molecule and that of the crystalline molecule. Irrespective of the manner in which one conceives of the molecular structure built around a carbon atom, a methane derivative whose four hydrogen atoms are replaced by four identical residues R situated at the same distance from that central atom should yield, if it underwent crystallization, a compound belonging to the cubic system; yet experience shows this not to be the case. These findings run counter to Mr. Wyrouboff's theory.

"In order for the tetrahedron hypothesis to be confirmed, it suffices to demonstrate that the chemical molecule, on passing to the state of a crystalline molecule, never has symmetry inferior to that derived from the consideration of the tetrahedron."

In fact, modern crystallography has amply shown that, while symmetrical (or achiral) molecules can sometimes crystallize in chiral arrangements (as in the case of potassium chlorate, which we have already met), chiral molecules, for their part, cannot crystallize into anything other than chiral systems.

Wyrouboff never capitulated. As late as in 1901 he wrote, in the *Bulletin de la Société française de la minéralogie* (Bulletin of the French Mineralogy Society):

"Are the two rotatory powers [that of crystals and that of solutions] analogous phenomena or do they stem from the same cause? Enticed by the ingenious metaphysics of 'stereochemistry,' the vast majority do not hesitate to answer in the negative" (and so on).

This was at least a recognition by our stubborn positivist that he had ended up belonging to the minority and that the *ingenious metaphysics* he rejected had triumphed.

All these rear-guard struggles could not prevent the development and extension of "three-dimensional chemistry": asymmetric carbon is only a special case of the "Pasteur principle," according to which any molecule whose image is not superimposable on it can actually exist in its two enantiomeric forms.

EXTENSION OF THE "PASTEUR PRINCIPLE"

Today, with the benefit of hindsight, we find it almost natural that after his stroke of genius in the field of combinations in which carbon is what supports chirality, Le Bel immediately thought that nitrogen might play the same role, since it too is an element that can be responsible for an arrangement of atoms that is not superimposable on its mirror image. This is the case, in particular, when the nitrogen is involved in a positively charged ammonium ion that is itself neutralized by any negative ion.

In 1891 Le Bel, after numerous unsuccessful trials in which he subjected solutions of certain ammonium salts to the action of molds (we shall come back to this method further on), had succeeded in observing deflections of polarized light. These solutions were definitely *levorotatory* (that is, they rotated the plane of light to the left). But it was not, in fact, until 1899 that the English chemists W. J. Pope and S. J. Peachey succeeded in separating the two enantiomers of a salt of the same type: this was the first example of the resolution of a combination whose chirality was not due to a carbon atom.

Since that time, many substances having an asymmetric atom belonging to elements other than carbon or nitrogen have been successfully resolved. These new structures, involving silicon or arsenic, for example, are in the direct line of descent from the discoveries of Le Bel and van't Hoff and their tetrahedron with differentiated apices. Manifestly, we need not give a more detailed inventory of them here.

Yet Pasteur's ideas begin to take on their full general value especially at the point where chemists become interested in molecules whose chirality is no longer associated with the existence of this type of special arrangement around a given atom, but stems from the specific character of the molecular object considered as a whole.

Van't Hoff himself, in his brochure *Chemistry in Space: Ten Years in the History of a Theory* (1887), had already envisaged such a possibility. Allenes, or allenic derivatives, are characterized by three contiguous carbon atoms connected by double bonds $>C = C = C<$ in such a way that, if one bond is situated in one plane, the other is in a plane perpendicular to the first. As a result, if the two substituents attached to the two outside carbons are different, we obtain a molecule whose "double" is not superimposable on it. The resolution of a molecule of this type, long considered unrealizable from the practical standpoint, was finally achieved by P. Maitland and W. H. Mills of England, more than sixty years after van't Hoff's prediction.

Another possibility quite analogous to that of the allenes occurs in the case of *atropoismers* (molecules that prevent rotation), of which derivatives of *biphenyl*, the hydrocarbon used to preserve citrus, provided the first examples. Imagine two benzene rings held together by a single bond. In the general case, these two plane hexagons can rotate freely about the axis that unites them; but if, by some artifice, we prevent this movement (by creating some sort of catch or stop on the benzene rings that prevents them from rotating freely), we produce a molecular object that looks very much like a two-bladed propeller. These molecular helices, like the propellers on boats, may be oriented in either of two directions. The first time an acid of the biphenyl series was resolved was in 1922, at the hands of the English scientists G. H. Christie and J. Kenner.

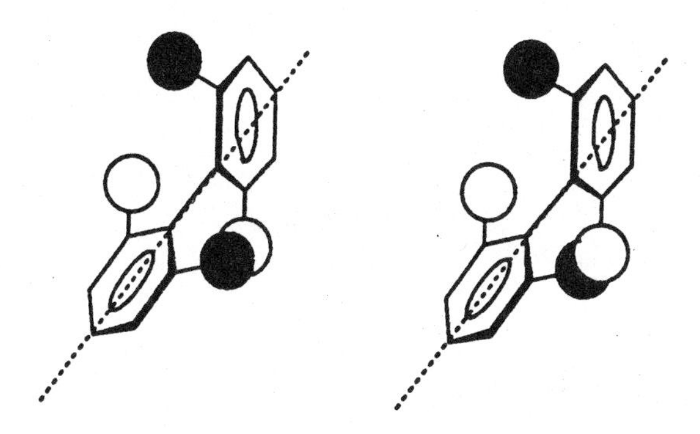

Now, isn't preparing and resolving these molecules made from coupled benzene rings that have one of the sides of their hexagon in common and are called, naturally, *helicenes*, quite simply a very elegant way to give material existence to the Pasteurian metaphor of the two spiral staircases, one turning to the right and the other to the left? The resolving of a hydrocarbon made up of *six* of these benzene rings, whose extremities must necessarily overlap either above or below (the whole molecule can no longer fit in a single plane) was achieved in 1956 by the Americans M. S. Newman and D. Lednicer.

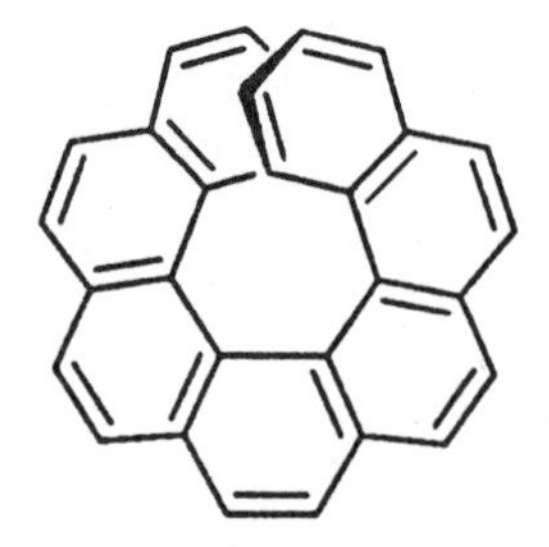

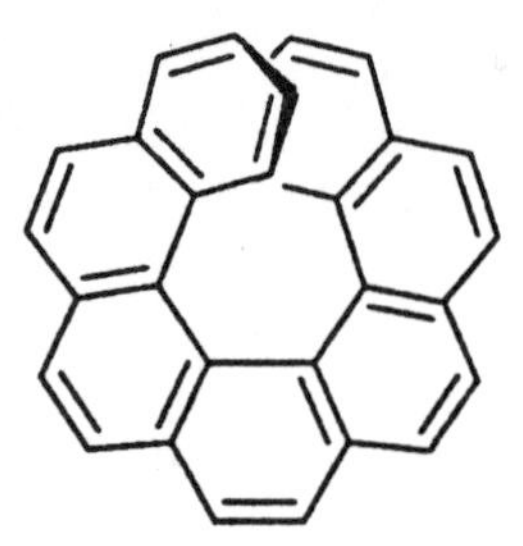

It was no doubt Alfred Werner, the inventor of the chemistry of mineral "complexes," who was the first, in 1912, to speak of the "Pasteur principle." Werner, who was born in Mulhouse on the eve of the 1870 war, went to live in Zürich when he was 16 years old and never moved from there. A professor at the Federal Polytechnic Institute and a lover of good white wines and peaceful ways, he had, in addition, a certain genius. He showed that some metals, such as cobalt and platinum, could give rise to combinations in which the chirality was organized around an *octahedral* structure. To give an example, without going too deeply into the details: If, in a regular octahedron having at its center a metal atom (M: platinum, cobalt, etc.), two neighboring apices are connected by means of a suitable organic molecule, then by repeating this operation three times one can obtain yet another resolvable "complex."

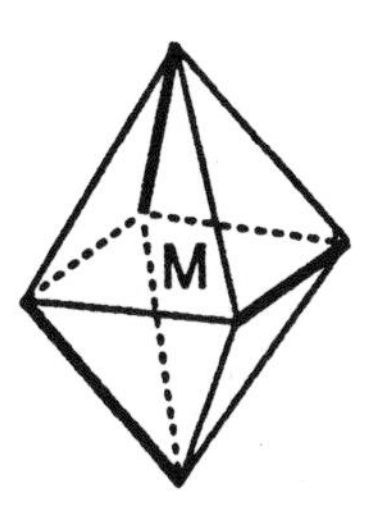

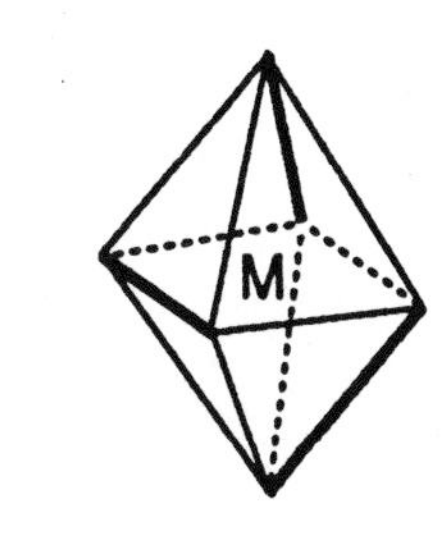

At this point in our review, now that we have taken a look at Pasteur's work and its consequences, we can perhaps go back for a moment to the crystals and the peculiar features of their mixtures whose separation was the starting point of our story. This digression will help us better to understand the conditions in which Pasteur's adventure took place and to pinpoint the difficulties and prerequisites of the process of resolving racemic compounds, or racemates, into their two constituents, a process that is so important from the practical standpoint.

BACK TO THE CRYSTAL AND THE DIFFERENT VARIETIES OF RACEMIC COMPOUNDS

Perhaps, however, before talking about crystals again, we should explain why we continue to take such a dogged interest in them. First of all it is because, in the eyes of the chemist concerned with molecular chirality, the solid state occupies pride of place, as compared with the liquid or gaseous state. As we shall see further on, with regard to the pharmaceutical industry in particular, gaining access to chiral molecules by resolving synthetic racemates remains a fundamental practical problem (and will no doubt become more and more so). Now, if we leave aside processes coming under biotechnology, such separations are based *almost exclusively on crystallization.*

The separation of enantiomer mixtures in the liquid state using chromatographic methods, though it can be done for the purposes of analysis or on a small scale, is rarely practicable for large quantities. It will be recalled that chromatography takes advantage of differences in molecular affinity on the surface of certain solids (called *adsorbents*) or at the interface of liquid films, due to the fact that such differences are reflected in the rates at which the molecules in question, carried along by a liquid or gaseous stream, "filter" through these special substances. In the case that concerns us, separation of enantiomers is obviously possible only if one uses adsorbents that are chiral themselves.

Resolution of mixtures of optical antipodes by distillation is impossible: the physical properties of the enantiomers are virtually indistinguishable, with the exception of their rotatory power. Crystals, which were instrumental in opening our eyes to the field that concerns us, thus remain for us a preferred object of study.

Pasteur, as we have seen, had observed the spontaneous resolution of the double tartrate of sodium and ammonium. He had also noted that this spontaneous separation did not take place in many other cases. This meant that there existed at least two species of racemate with different properties. Of what did this difference consist? Here again, the answer to the question will come from another discipline, located on the fringes of chemistry and crystallography: thermodynamics.

Today we know that there exist, in the crystalline state, *three* solid species that consist of a mixture of equal parts of two enantiomeric molecules.

In the first, known as *conglomerate*, the two enantiomers coexist and retain their individual characteristics. From the point of view of thermodynamics, it is as though one of them were unaware of the presence of the other. Thus, Meyerhoffer, a disciple of van't Hoff who died a premature death, demonstrated that the solubility of a racemate of this type was twice that of a pure enantiomer. This means that when a mixture of the two enantiomers in any proportion is crystallized, it is always the more abundant one that can be recovered in pure crystal form, while the equimolecular mixture of the two remains in solution.

In the case of organic compounds that have a melting point (which is not always the case with mineral salts), the melting point of this conglomerate can be computed in a simple way from the physical constants (temperature and heat of fusion) of the enantiomers themselves, as is generally the case (with only a few exceptions) with any mixture of two chemical individuals whatsoever having no particular relationship to one another. Thermodynamics explains to us why the melting point of such a conglomerate is generally about twenty degrees Celsius below that of the pure enantiomers of which it is composed.

This variety of racemate, as we have said, is relatively rare: The sodium ammonium tartrate whose spontaneous resolution Pasteur was the first to note belongs to this type.

Those referred to as *true* racemates, which make up the second crystalline variety of equimolecular mixtures of enantiomers, are the most common by far. In these cases, the two molecules of opposite rotatory power are combined, so to speak, in the same crystal lattice. In the present state of the art, the properties of this second racemic species are not yet predictable from those of the molecules that constitute them: the solubility of the racemates may be greater than, equal to, or lower than that of the related pure enantiomers. The same is true of their melting point. These differences obviously reflect the varying degrees of stability of the crystal lattice to which the two enantiomeric molecules belong and the solidity of the bonds that unite them.

If, as may sometimes be necessary, we wish to recrystallize an incompletely resolved enantiomer (namely, one containing a certain percentage of its "double," hidden, in this case, in the racemic compound), we may observe, depending on the case, that it becomes poorer or richer in undesirable isomer content. These possible outcomes, which may or may not be favorable, can now be predicted—but that is a more difficult field, into which the layman need not be drawn. It will suffice to note that this field has now largely lost most of the mysteries that long intrigued the chemists of the last century.

The third and last species of crystallized racemate is by far the rarest. Whereas in true racemic compounds, the association of the two enantiomers in equal quantities constitutes a clearly defined new species, that is not the case with *pseudoracemates*. Here, for a variety of reasons, the two enantiomeric molecules seem to resemble one another so much that they can be accommodated by the same crystal lattice, to the point of being merged in it and becoming interchangeable—dissolving in one another, so to speak. Here, in a special case, we find an example of *isomorphism*, which we had been led to by Mitscherlich. This possibility of *solid solutions* between enantiomers, which we shall simply mention, relates chiefly to molecules which, owing to their geometry, resemble spheres (and thus differ little from their mirror images). The two antipodes of *camphor*, for example, which are typical of *globular* molecules, belong to this special class of racemic compounds.

The industrial practice of performing such resolutions can and obviously must take into account and benefit from the foregoing observations. Thus, however rare it may be, the existence of a conglomerate, when one is lucky enough to encounter it, may give rise to the separation of enantiomers in the absence of any auxiliary chemical agent. When this happens, it is possible, under certain specific conditions, following the technique of *resolution by entrainment*, to work with the different crystallization rates of the two compounds to be separated and collect first the one, then the other. Among other cases in which this technique could be used, we might mention that for several years it has been applied to good advantage in the industrial preparation of *chloromycetin*, a synthetic antibiotic only one of whose two enantiomers is therapeutically active.

More generally, however, such separation of isomers is achieved in practice through the formation and fractional recrystallization of salts combining the substance to be resolved with a plentiful, cheap chiral molecule, usually one found in nature. Here, Pasteur's observations regarding the separation of cinchonine tartrate remain exemplary.

To resolve basic molecules, recourse is had to acids, either natural (which are abundant and inexpensive, such as tartaric acid) or readily obtainable from natural products (such as certain camphor derivatives). Thus certain tranquillizers or certain appetite-suppressing preparations are similarly resolved in the form of tartrates. Conversely, when acids are to be separated, use is made of bases, again natural and readily obtainable ones, such as cinchonine or quinine. This is the case with some anti-inflammatories.

HOW CAN WE DESCRIBE AND KNOW ABSOLUTE CONFIGURATIONS ?

Let us return once again to tartaric acid and agree to write its stereochemical formula as follows:

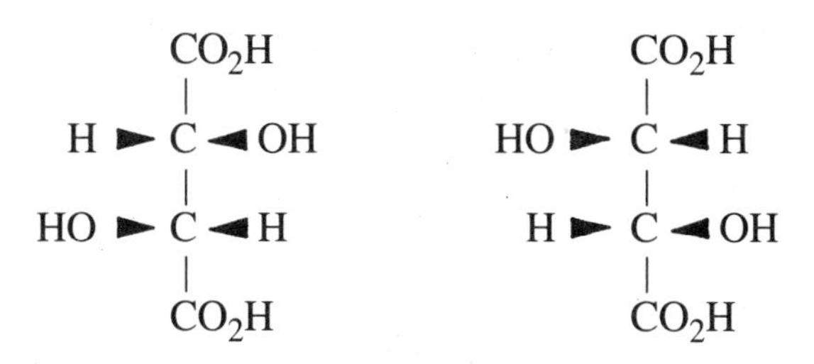

The two central asymmetric carbons are shown one beneath the other in the plane of the page, while the two substituents (H and OH) that surround them "come out" of the page toward the reader. The two carbons located at both ends of the molecule, on the other hand, project away from the reader.

Of these two possible formulas, which is the one that matches the real three-dimensional arrangement of the atoms that make up *dextrorotatory* natural tartaric acid (namely, rotating the plane of polarized light to the right)? What is their *absolute configuration*, as it is called, and how can we know it? Again it was crystallography that, in 1951, provided the solution to this question raised by Le Bel and van't Hoff seventy-five years earlier.

Thanks to our knowledge of reaction mechanisms, chemical procedures enable us to determine the *relative* configurations of several asymmetric carbon atoms present in the *same* molecule. To take a simple example, it is obvious that the two asymmetric carbon atoms of "inactive" (or *meso*) tartaric acid have *opposite* configurations since juxtaposing them results in a molecule that has no effect on polarized light, by a sort of compensation process. This type of deduction is also possible when one tries to establish a stereochemical relationship between *two* different molecules. We have seen, for example, that it is possible to transform aspartic acid into malic acid. If we knew the relative positions of the four substituents that surround the asymmetric carbon atom in the first, we could determine the positions that they occupy in the second. Chemists pride themselves on certain acrobatics, known only to themselves, whereby from a single compound of known configuration they can establish the configurations of all other chiral compounds that can be derived from it. The problem of their absolute configuration, however, is not resolved at all so long as we do not know that of at least one of them.

We have seen that chemists—and they are not the only ones—do not shrink from giving a name to something they do not yet know: They know how to baptize a substance even before they have prepared it. To be sure, problems of nomenclature lie right at the heart of their preoccupations: without it, they cannot conveniently communicate. We should not be surprised, therefore, if they have taken pains to invent drawing and writing conventions to enable them to represent an absolute configuration even before they were sure they would ever be able to know one. It was especially in the area of sugars, of which we have had a glimpse of the complexity, that these problems began to crop up. This research is dominated by the powerful personality of Emil Fischer (1852-1919), the German who was awarded the second Nobel Prize for chemistry in 1902, the year after van't Hoff became the first to be so honored.

On the occasion of the naming of the various isomeric sugars known in 1891, Fischer proposed a method that was long in use and is still used in certain cases.

Fischer considered a particular molecule (*glyceraldehyde*) from which he was able to synthesize certain natural sugars. He wrote the space formula of this reference compound in accordance with the conventions which we applied above in the tartaric acid formulas. He then decided to represent his glyceraldehyde rotating polarized light to the *right* by a formula having its OH at the top *right*, and to assign to it the configuration capital D. The mirror image of this formula of course had to be L, the formula of the levorotatory compound. It must be clearly understood that this decision to associate this nomenclature D and L with the sign of the rotatory power—d and l or (+) and (–)—*was purely arbitrary and conventional.* In fact, Fischer had one chance out of two of making a mistake. The subsequent developments will show that he had drawn the winning number.

It must also be noted that although there was originally a relationship between the choice of these capital letters and rotatory power, certain chiral substances designated in accordance with these rules (which were subsequently refined and adjusted) may belong to the "D series" despite the fact that they possess left-handed rotatory power and vice-versa. In any event, this system of nomenclature, which still exists today in a few chemical catalogs, involves many difficulties of application, which we need not go into here. For some time now, it has been giving way to the system proposed jointly in 1956 by R. S. Cahn and Sir Christopher Ingold, both English, and Vladimir Prelog, a Swiss.

This system, which has very broad applications, has largely been adopted by the community of chemists interested in stereochemistry. We can sum it up in one general rule and a few subrules, describing in images the procedure whereby the absolute configuration of an asymmetric carbon atom A can be designated unambiguously. The four substituents are classified according to an order of priority established by the subrules (a > b > c > d), and a, b and c are placed on a steering wheel, with the atom d (the "smallest" one) located on the shaft. If, on turning the steering wheel in the direction a → b → c, the driver goes toward the right, the asymmetric atom is R (from the Latin *rectus*, right). It is S (*sinister*, left) if the driver goes to the left. We shall not go into the technical details of these subrules for establishing the order of classification of a, b, c and d; a single example will suffice to give an idea of the process. Chlorine, bromine and fluorine are classified according to their *atomic numbers*, a higher number coming before a lower one: $^{35}Br > ^{17}Cl > ^{9}F$.

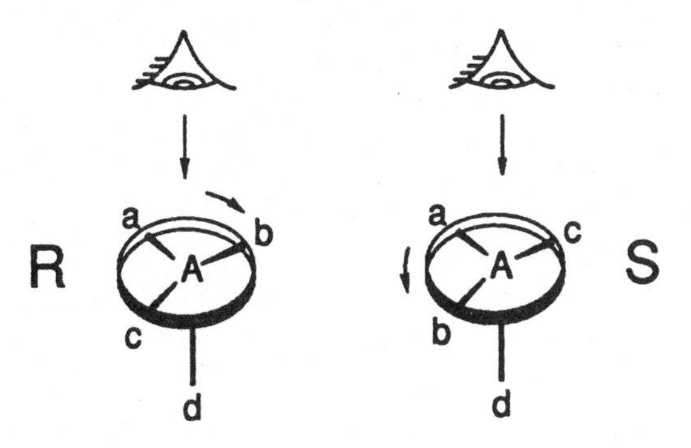

But how can we ascertain the real configuration of a real molecule, characterized by the sign of its rotatory power? How do we know, in fact, whether it is R or S?

We do know, since Sir William Henry Bragg (1862-1942) and his son Sir William Laurence Bragg (1890-1971), both Nobel Prize winners in 1915, that the diffraction of x-rays by a crystal enables us to locate the atoms that constitute it (though it does not specify the nature of the element to which they belong). By means of a few sophisticated calculations, crystal chemists can construct a good three-dimensional image of the molecules contained in the crystal. However, since x-rays are not dissymmetric, they do not make any distinction between a right-handed and a left-handed molecule.

Imagine, however, a molecule, which we shall represent schematically by A–B, and its mirror image, represented by B–A. Radiation strikes points (atoms) A and B in this first arrangement. The diffracted waves will produce an interference figure at point I, with, for example, a distance A-I greater than B-I. If we invert A and B, the path of the wave diffracted by B (B-I) will now be longer than A-I. The interference figures at I, however, will remain unchanged, due to the fact that the difference between the distances A-I and B-I has remained the same. How can we distinguish the beam diffracted by A from the one diffracted by B?

If, at A (for example), we can cause a phase lag in the diffraction process, the already slower radiation diffracted by A (the trajectory A-I being longer than B-I) will be retarded still further by the phase lag and the phase difference between the two interfering radiations will be *increased.* With the positions of B and A "inverted," the radiation whose phase was delayed at A was initially more rapid (since in this case A-I was shorter than B-I) and the phase difference between the two radiations will *diminish.* In short, the two interference patterns will not be the same, and it is possible to distinguish between the one corresponding to the arrangement A–B and that corresponding to B–A, provided that we know which atom is causing the disturbance.

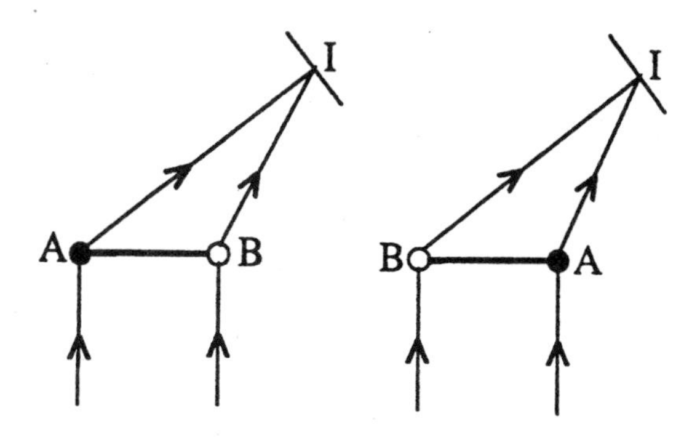

The way one introduces a phase delay in the diffraction at point A (and not at B) is to use a beam of x-rays having a wave length close to the absorption range (for x-rays) of atom A. In the original experiment of the Dutchman Johannes Martin Bijvoet (1892-1980) and his collaborators, who invented this method of determining the absolute configuration of natural tartaric acid, the double tartrate of sodium and rubidium employed was irradiated with x-rays produced by zirconium (K-line). Since rubidium (Rb) [and not sodium (Na)] produces retardation of the phase of the refracted radiation, it finally became possible to deduce that tartaric acid had the *absolute configuration* illustrated by the following figure:

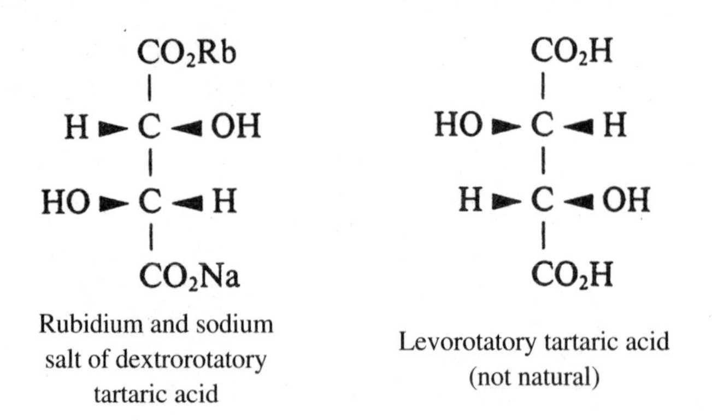

Rubidium and sodium salt of dextrorotatory tartaric acid

Levorotatory tartaric acid (not natural)

This method, extended to compounds containing a bromine atom irradiated by specified uranium radiation, has been widely used since that time for determining other absolute configurations. Thus, once we know the rules of *asymmetric induction*, of which we shall speak, and the mechanisms at work, it becomes possible, by chemical means, to deduce from these "first generation" absolute configurations virtually all those of the compounds one wishes to know.

PROPAGATING CHIRALITY

Today, chemistry, left to its own devices, does not know how to create *ex nihilo* a molecule exhibiting rotatory power. We have just seen that the means it employs to achieve this goal involves extensive use of the chiral materials generously offered by Nature. Actually, to be more precise, there exist a few cases in which chemists have succeeded in obtaining optically active products through recourse to reactions involving nothing more than the dissymmetry of a polarized light. But the rare cases in which this result (which we shall examine further on) has been achieved have, one might say, only a symbolic value: They prove that nothing stands in the way of it from the theoretical standpoint. But let us remain in the field of current practice, at least for the time being.

Chemists are able to resolve or take advantage of the spontaneous resolution of the molecules they study or transform. Above all, they know how to put to work this initially acquired chirality to create other centers of asymmetry, where they had been only potential, or transfer chirality from one molecule to another. In short, while they may not know (very well) how to create chirality, they do know how to propagate it.

The essential features of these procedures can nearly always be summarized as follows: An asymmetric center of a given molecule can determine the configuration of a new asymmetric center created by a reaction that affects an atom in its vicinity. This phenomenon, known as *asymmetric induction*, can give rise to multiple variants, which need not be listed here. Let us simply note that over the past few decades, the invention, use and mastery of these *stereoselective* reactions (that is, reactions that precisely control the configuration of the molecules obtained) have occupied the limelight for many organic chemists.

Two essential factors come into play in determining the *purity* of the products obtained by means of these reactions and the *efficiency* of the process. In one scenario, such reactions yield two stereoisomers with different *stabilities*. When these compounds can be transformed back and forth into one another, a situation of equilibrium may come into being, resulting in a mixture in which the proportion of each of the products depends on their relative energies. We then say that the stereochemical result is *thermodynamically controlled*. In the other scenario, the result of the reaction is *kinetically controlled*: in other words, the dominant product has developed *more rapidly* than the minor stereoisomer. This may be the case particularly when, owing to hindrance around the atom being attacked, the reagent can approach more easily from one side than from the other.

These problems of stereoselectivity, or in other words, the controlled elaboration of the architecture of the reaction products, may seem quite abstract and academic to the uninitiate; yet the fact is that they are often of enormous practical importance. It is not unusual in the pharmaceutical or phytosanitary industry for the synthesis of a given chemical agent to require a series of many, complex transformations. The synthesis of some cortisone analogues, for example, requires several dozen stages, each involving the obtainment and purification of a very specific intermediate product. Running through the "production line" that goes from the raw material to the drug or pesticide ultimately offered to the consumer often takes several weeks, and sometimes several months. In this type of industrial activity, the question of the yields obtained at each stage is obviously of prime importance. In a manufacturing "run" that includes, for example, forty stages—which might be the case in the preparation of certain particularly complicated hormones—and is aimed at producing one kilogram of final product, an average yield of 80% at each stage would require 7.5 tons of raw materials to start with, whereas only 68 kilograms would be required if the yield could be brought up to 90%.

In the case of the optically active molecules that concern us here, production constraints lead one to contemplate at the same time a veritable *economy of chirality*, by which we mean a manufacturing strategy that takes into account the need not to waste chirality, a valuable "commodity" which, as we have already said, can come only from a natural (vegetable or animal) source or—in a somewhat roundabout way—from a more or less laborious resolution process. We can give an idea of the true nature of these problems by taking a look at a concrete example.

"Dalmatian insect powder," or "pyrethrum flowers," is a traditional insecticide obtained from the dried flowers of certain chrysanthemum varieties. It owes this activity to specific substances which chemists have studied and the structure of which was elucidated in the 1920s. These interesting molecules derived from chrysanthemic acid were subsequently transformed and modified in varying degrees, giving rise to a family of artificial compounds known as *pyrethroids*.

During the 1960s and 1970s the chemists of the British firm Imperial Chemical Industries (ICI) described how they obtained pyrethroids having insecticidal properties that were remarkable both for their resistance to direct sunlight and their low toxicity to mammals. *Deltamethrin*, for example, prepared industrially by the Roussel-Uclaf company, is roughly 100,000 times more toxic to insects than to rats.

These synthetic insecticides, very closely related to the natural products, are very often marketed in the optically active state. This is the case, in particular, with *deltamethrin*, whose molecule includes three asymmetric carbons. It is thus capable of existing in the form of $2^3 = 8$ isomers (or four racemic pairs). If we number these three carbons C_1, C_2 and C_3 and designate them in accordance with the rules of R and S nomenclature, we can predict the following combinations (which have actually been prepared):

C_1	R	S	R	S	R	S	S	R
C_2	R	S	R	S	S	R	R	S
C_3	R	S	S	R	S	R	S	R

Six of these eight stereoisomers have no insecticidal activity. Thus one can understand why it is essential to manufacture stereoselectively the most advantageous product, *deltamethrin*, which, among these different possibilities, matches the structure R R S.

Chrysanthemic acid, the initial product used in its synthesis, is prepared in the racemic state in an initial stage. As of this stage it is immediately separated into its two antipodes, by means of a reagent that is also obtained by resolution by entrainment during the manufacture of an antibiotic marketed by the company.

Generally speaking, it is very important in "multistage" industrial synthesis of this type for the resolution processes to take place as early as possible in the chain of anticipated reactions, for those processes in fact result in the rejection of half of the racemic preparation used: The fewer transformations the useless enantiomer has undergone, the smaller its "value added" will be. However, even if the cost of the discarded antipode is low, it is always advantageous to provide for its reuse. This presupposes, for example, the racemization of the unusable isomer, which can than be fed back into the production line in the form of its optical antipode.

We, at this point, cannot go into the technical details of the subsequent stages of preparation of this insecticide. Let us simply add, continuing to use the symbolic nomenclature mentioned above, that "at the end of the road" the chemist winds up with two isomers, one of the "bad" ones (RRR) and the "good" one (RRS). As luck would have it, the latter is the more stable and, when crystallization takes place, it is deltamethrin that is isolated, thanks to its lower solubility.

III

CHIRALITY AND LIVING THINGS

PASTEUR AND MOLDS

Who can express it better than Pasteur himself? "We are arriving at last at a final chapter that is sure to be no less interesting than any of the preceding ones, for it will provide us with clear evidence of the influence of dissymmetry in the phenomena of life."

For again it is Pasteur who has given us the keys to another part of the domain that we will investigate now: How do living things react to a chiral molecule? How and why are the molecules that make up living organisms chiral themselves?

"It had been known for a long time, thanks to the observation of a German chemical manufacturer, that impure calcium tartrate from factories, contaminated with animal matter and left under water in summer, could ferment and yield various products.

"Given this, I fermented ordinary dextro-ammonium tartrate in the following manner: I took the very pure, crystallized salt and dissolved it, adding to the liquid a very clear solution of albuminoid matter. One gram of dry albuminoid matter suffices for 100 grams of tartrate. Very often it happens that the liquid, placed in an oven, ferments spontaneously.

"So far, nothing special: a fermenting tartrate—something familiar to all of us.

"But let us apply this mode of fermentation to the ammonia racemate, and under the above conditions it ferments. The same yeast is deposited. Everything seems to indicate that things are taking place exactly as in the case of the dextrorotatory tartrate. However, if we follow the progress of the operation with the polarization apparatus, we quickly recognize profound differences between the two operations. The originally inactive liquid has appreciable left rotatory power that gradually increases until it reaches a peak. The fermentation is suspended. There is no longer any trace of dextrorotatory acid in the liquor, which, on being evaporated and mixed with an equal volume of alcohol, immediately yields a fine crystallization of levo-ammonium tartrate."

This observation is often cited as the third of the methods invented by Pasteur in 1857 to resolve racemic acid, after having taken advantage of the existence of a conglomerate, then of diastereoisomeric salts. To be more precise: He shows us—and this is no doubt still more important—that *Penicillium glaucum* knows perfectly well how to distinguish between the two antipodal tartaric acids, causing one to disappear by feeding on it and sparing the other. It must be recognized that this is less a matter of separation than of selective destruction.

This resolution method, which sacrifices the "double" of the molecule obtained (or more exactly, one-half of the racemic compound), exhibits a further major drawback: Indeed, the chemist is not the master of a choice that stems from the selective taste of a microorganism. It sometimes prefers to feed on the enantiomer that the chemist needs. Nevertheless, Le Bel still managed, as early as 1891, to use this process to obtain an optically active alcohol (specifically, *amyl* alcohol) in a case in which, owing to the nature of this type of compound, being neither acid nor base, it was not possible to prepare a salt directly, as in the case of the acids resolved by Pasteur.

As one would suspect, the discovery of this asymmetric destruction of which molds are capable opened new chapters in the budding field of biochemistry.

In the early 1890s Emil Fischer had undertaken his remarkable work on sugars, their constitution and their synthesis, of which we have already spoken. He had prepared racemic fructose, mannose, glucose and galactose, among others. By subjecting these substances to the action of baker's yeast, he had found that it was possible to ferment only the "natural" sugars present in the product inactive with respect to polarized light and obtained by synthesis. They were destroyed, while their nonnatural enantiomers remained intact, as proven by the rotatory power that appeared in the reaction liquid. He subsequently ascertained that all the natural sugars known at the time that were subjected to the action of the same microorganisms exhibited behaviors that were closely dependent on their stereochemistry: some were fermentable, others not. "Among the multiple geometric forms" [of these sugars], he concluded, "there are only a few that satisfy the needs of the cells." "Beyond the shadow of a doubt," he continued, "similar observations will be possible for other microorganisms and for other groups of organic molecules. ... There are certainly numerous chemical processes that go on within living organisms and are sensitive to molecular geometry. One can assume that yeast cells and the asymmetric agents they contain can only ferment sugars that are not too far from the geometry of grape sugar [dextrose]."

Continuing his research along these lines, Fischer extended his observations to other aspects of stereoselectivity, this time using not whole microorganisms, but the *enzymes* they secrete, long recognized as being responsible for certain extraordinary specific chemical activities.

It had been known since 1833, in fact, thanks to the studies of the French chemists Anselme Payen (1795-1871) and Jean-François Persoz (1805-1868), that there existed in living organisms "catalysts" whose presence permitted chemical reactions that would not take place without them. Later, in 1897, the German biochemist Eduard Buchner (1860-1917, Nobel Prize in 1917) had shown that these enzymes (or diastases) maintained their catalytic activity even when they were separated from the cellular organizations that housed them—in short, that apart from their complexity, these enzymes were like any other chemical molecules which had not been imprinted with any hypothetical "vital force."

Fischer, for his part, confirmed that, as far as the configuration of the substances they attacked was concerned, these enzymes were just as selective as the yeasts or other lower organisms from which they were extracted. *Invertin* (which splits cane sugar into glucose and levulose) and *emulsin* (which can be extracted from sweet almonds), for example, are *proteins* which, like other proteins, exhibit an asymmetric molecular structure. Their specific activity on certain complex sugars (such as the *glycosides*) was proof for Fischer that the chemical processes they set in motion were founded on perfect matching of the geometry of enzymes to that of their substrates. "To use an image," he concluded in a formula that has become famous, "in order for a reaction to take place, the enzyme and the glycoside must fit together *like a key in a lock*."

The inexhaustible wealth of possibilities of enzymatic chemistry alone would merit lengthy treatment. We shall simply illustrate it with two examples that relate directly to our subject: the obtainment of chiral substances in the pure state.

There exist chemical substances which have no asymmetric carbons but whose structure is such that a very simple reaction often suffices to create the possibility of optical isomerism in the product of transformation. These structures are known as *prochiral* (or, more precisely, the atom directly concerned that will become asymmetric is referred to as *prochiral*). Some enzymes are perfectly capable of effecting this type of reaction and, for example, fixing a hydrogen molecule on an achiral ketone in order to produce a chiral alcohol in the form of a single enantiomer in the pure state.

In another type of reaction, certain *esterases* are able to split a racemic ester into its two potential constituents (an alcohol and an acid) stereospecifically: only one of the enantiomers is transformed, the other remaining intact. This operation clearly reminds one of the historic "resolution" of tartaric acid by molds—a more profitable operation, to be sure, but not one that would have surprised Pasteur.

BIOLOGICAL RECEPTORS AND CHIRALITY

In 1886, A. Piutti had discovered that *asparagine* extracted from asparagus (an aspartic acid derivative about which we have already spoken) also had its "double": it was able to exist in the form of two spontaneously separable optical antipodes. This had enabled Piutti to make a surprising observation: "While ordinary asparagine has a nondescript flavor, dextrorotatory asparagine is definitely sweet to the taste."

Pasteur, who was unquestionably still on the scene, deemed it appropriate to comment on this finding after communicating the note of his young colleague to the Academy:

"What is the reason for this considerable difference in the flavor of the two asparagines? One might perhaps assume the existence of a very special isomerism. I think not. I would be very inclined to believe, on the contrary, that this physiological phenomenon must be viewed in connection with another: namely that, while two opposite dissymmetric bodies may, in their combinations with inactive bodies, exhibit chemical and physical properties that are absolutely similar and even identical, these same opposite dissymmetric bodies yield entirely different combinations of properties when they unite with bodies that are themselves dissymmetric and active with respect to polarized light.

"The dissymmetric active body that comes into play in the nerve-transmitted impression reflected in a taste that is sweet in one case and practically insipid in the other is, as I see it, nothing but the nervous material itself, a dissymmetric material like all the primordial substances of life: albumin, fibrin, gelatin, etc.

"But, you will say, how is it that no differences in taste have yet been found in opposite right-handed and left-handed bodies?

"That is not a fundamental objection. Furthermore, perhaps no one has ever had the idea of making such taste comparisons. Now that attention has been called by what I have just said to these peculiarities of great importance, maybe things will change—at least I hope they will."

Piutti's observation was confirmed fairly soon and extended to other pairs of enantiomeric amino acids as soon as it was possible to prepare them. During the first years of this century, Emil Fischer and his co-workers methodically tasted the synthetic amino acids they obtained. The results of these austere tastings can be summarized in a few words: *Nonnatural* amino acids belonging to the D series (alanine, serine, valine, leucine, isoleucine, histidine and asparagine) are all more or less *sweet*, whereas their *natural* counterparts are either less sweet or relatively tasteless or even *bitter*.

Two amino acids can be combined to produce a *dipeptide*: *Aspartame* is one of the best known, at least to those who like their coffee sweet but want to keep trim. This beneficent molecule combines aspartic acid and phenylalanine, both of which belong to the *natural* L series. Masur and his fellow workers, to whom we owe the (chance) discovery, in 1969, of the sweetening properties of this LL dipeptide (that is, made up of two L amino acids), were no less surprised to find out that the DD enantiomer (namely, the one derived from the *nonnatural* series) was bitter. Logically, given the tastes of the amino acids that make up the two antipodes, one might have expected the contrary. The same unexpected bitterness is also found in the DL and LD isomers, which they took the pains to prepare.

With regard to sugars, our curiosity concerning the taste of the isomers will not be so easily satisfied. This is due to the fact that the nonnatural enantiomers, which are very complicated to obtain free of any trace of the other enantiomer, have often only been glimpsed and have no doubt never been tasted for the purposes of scientific examination. We know, however, that in the case of isomers having several asymmetric carbons it suffices to reverse the configuration of only one of them in order to effect a profound change in gustatory properties. In the specific case of D-glucose and D-galactose, which differ in only one of their four asymmetric carbon atoms (DDLD and DLLD), the first is twice as sweet as the second.

Karl von Frisch (1886-1982, Nobel Prize in 1973), who discovered the language of bees, was also interested in the subtlety of their sense of taste. He demonstrated, by means of a number of ingenious tests, that bees could distinguish between the isomeric sugars offered to them as food. To them, D-fructose tastes good, while D-sorbose does not, and yet here again these two sugars differ in the configuration of only one of their asymmetric carbon atoms. The same is true of L-rhamnose and L-fructose: The bees gather nectar from the former only.

The role of molecular chirality in the perception of odors is the focus of research and controversies that are still current. Robert Hamilton Wright, in his theory of vibrations (1963), considers that optical isomerism has but little importance.

To R. H. Wright, the mechanism of olfaction is essentially physical in nature; the chemical interactions between the scented molecule and the receptor-detector, he feels, play only a secondary role. The odor of a molecule is determined by the frequency of its vibrations in the far infrared region (500-50 cm^{-1}), while its volatility or its solubility (in water and in fats) affect only the intensity of the perception.

John A. Amoore, on the other hand, stresses the importance of the form of the odoriferous molecule. He attributes primary importance to its spatial structure. His theory is not unlike that of E. Fischer, who likened the specific relationship that exists between an active compound and an enzyme to that of a key and a lock.

One must admit that the subject is still quite moot, and investigators run up against numerous obstacles. They have to do with the subjective nature of the more or less quantitative tests to which enantiomeric smells can be subjected: the "noses" belong to individuals, even if those individuals are professionals. The purity—the enantiomeric purity—of the chemical species examined, moreover, is not always easy to achieve or to determine. Last but not least, unlike the activity of the antipodes with respect to other biological receptors recognizing other molecules, the two enantiomers always have an odor. In this area there is no "all or nothing"; everything here is a matter of degree.

In one of the best-studied cases, using highly purified products, it was confirmed (in 1971) that levorotatory *carvone* (–) has an odor of green mint, while dextrorotatory carvone (+) has an odor of essence of caraway (an umbelliferous field plant to which it owes its name). For the small number of other examples that have also been carefully studied, the differences between enantiomers can be described only in very subtle terms: dextrorotatory *citronellol* (+) smells of oil of citronella, while the levorotatory form (–) has a more powerful odor reminiscent of essence of geranium.

It is perhaps legitimate to attribute the relative lack of difference in odor between enantiomers to the very great spatial resemblance of the substances concerned. The molecules in question, despite their volatility, are very small in size; they are often nearly spherical, thus not very dissymmetric; and if they are linear, their dissymmetry is often not very pronounced. In the case of the citronellols, for example, it is confined to the configuration of a single carbon atom, the nine others remaining superimposable. In the last analysis, the human sense of smell, as we know, is not the most sensitive. In other words, in relation to the complexity of the problems mentioned, the study of the olfactory sense of insects (which may possibly result in quantitative measurements and easier comparisons) seems, *a priori*, quite attractive.

A number of *pheromones* (from the Greek *pherein*, to carry, and the latter half of *hormone*, from the Greek *hormao*, to set in motion, stir up) which butterflies can "smell," sometimes from very far off, and which lead them to their females, are in fact chiral. Since the first studies by Adolf Butenandt *et al.* (1959) on the pheromone of the silkworm, *bombykol*, chemists have learned to isolate these sexual attractants, though they exist only as traces, and to determine their structure and synthesize them. By synthesis it was possible to obtain the two enantiomers of a large number of them, and we now know the sometimes considerable differences between the activity of the natural hormone and that of its antipode.

In 1974, three distinct groups of researchers working on the pheromones of different insects, made their contributions to this interesting problem. The Americans Robert G. Riley and Robert M. Silverstein prepared the two enantiomers of a pheromone isolated from *Atta texana* and showed that the dextrorotatory isomer was approximately 400 times more active than its antipode. At the same time, the Japanese team of Shingo Marumo was studying the two *disparlures*, hormones of the *gypsy moth* (*Porthetria dispar*), and showed that the dextrorotatory isomer was considerably more active than the other. Finally, after K. Mori and associates had prepared the two antipodes of *brevicomin*, they found that only the levorotatory isomer brought about the gregarious meeting of the pine sawflies that secrete it. It would be tedious and pointless to list all the pheromones prepared and tested over the past two decades. We shall simply retain the fact that this research, in which chemists and biologists work together, has shown that chirality plays an unquestionable and in fact decisive role in the recognition of the odor of a molecule by an insect, though that role is not always simple and not always clear. In particular, we observe that the effectiveness of pheromones may sometimes depend on the proportions in which the two antipodes are present. Curiously enough, variable quantities of one of them may enhance or diminish the efficacy of mixtures of them, depending on the case. We shall also retain the fact—need we be reminded?—that such research can be of the utmost practical interest in the effort to control insect pests.

The biological receptors about which we have just spoken react to molecular excitations coming from the *outside*. Those capable of recording chemical messages originating right *inside* a living organism, those responding to hormones secreted by the *endocrine glands* (Greek *endon*, within), are no less sensitive to chirality.

The synthesis of the first sex hormone obtained in its two antipodal forms was performed in 1939 by the American team of W. E. Bachmann *et al.* The hormone was *equilenin* (from the Latin *equus*, horse), isolated a few years earlier from mare urine. This hormone belongs to the family of *estrogens*, which control the onset of heat in animals and play a role in the menstrual cycle in women. It was found that levorotatory equilenin was only one-thirteenth as active as the natural hormone. Subsequently, the synthesis of the racemic form of *estrone*, the estrogenic hormone *par excellence*, revealed that this form was slightly more than half as potent as the natural hormone: in other words, its antipode was practically inactive. The Swiss team to whom we owe these interesting findings did not hesitate to conclude: "In accord with the findings of W. E. Bachmann regarding the isomers of equilenin, our own results confirm once again that nature always produces the most active of the sterically possible isomers." Such a statement, if made by a philosopher, would no doubt warrant lengthy commentaries. We might give any number of examples of the astonishing adaptation of biological receptors to the molecules that accept them and, one might say, vice versa, but those we have mentioned will no doubt suffice in order to show how interesting the problems are and what questions still remain open. But now it is time that we examine the response of the receptors of living things to chiral substances they have never yet seen. This is often the case of the medical preparations or chiral poisons that chemistry may propose to them.

RACEMIC DRUGS AND CHIRAL DRUGS

If one is concerned about classification, resolvable substances that have pharmaceutical properties can be divided into four groups, depending on the behavior of their constituent isomers in the organism as compared with that of their racemic mixture. Any differences among these different possibilities ought to have an immediate consequence: they determine the answer to the question that manufacturers, practitioners and users have the right or the duty to ask: Can one indiscriminately prescribe or take a "remedy" in either of its enantiomeric forms?

Obviously, therapeutic agents whose two enantiomers have the same properties present no particular problem. There is no advantage in separating them and no need to do so, especially when it has been confirmed that their transformation in the organism—their metabolism—follows an identical pattern. Those with enantiomers whose properties differ only in *quantitative* terms are hardly a problem any more: it will suffice to know that one is more effective than the other. In the extreme case, one of the antipodes has useful therapeutic activity, while the other is only a sort of "ballast," unnecessarily "filling out" the prescribed racemic compound by diluting the "right" enantiomer (though one can never be quite sure that the "wrong" one is entirely "neutral.")

Things are obviously not so simple in the last case, in which the two enantiomers of a given compound have *qualitatively* different properties. In fact, it is neither inconceivable nor uncommon for one and the same molecule to serve multiple purposes: as an antirheumatic, a pain-killer, an antipyretic, and so on. If it can be resolved, one of the enantiomers might, for example, be analgesic and the other more specifically antirheumatic. This type of preparation is unquestionably one in which resolution is a must, if only because patient sensitivity to these different effects ("side effects") are largely unpredictable.

A lengthy list of pharmacological data could be presented to illustrate the various possibilities we have just mentioned.

Let us just discuss one case that was quite spectacular in order to show that these distinctions are not mere schoolboy exercises.

We have not yet completely forgotten the tragedies caused by *thalidomide*, which had been recommended in the late 1950s as a potent sleep inducer and sedative, but, as it was found out too late, could cause disastrous fetal malformations in pregnant women. This "drug" was originally prescribed in racemic form. It was not until 1979 that tests of its two enantiomeric components revealed that *only the levorotatory S derivative was teratogenic*, while all three forms—the racemic, right-handed and left-handed forms—had roughly the same sedative activity.

This dramatic case was obviously an extreme one, but a great many other cases are known in which the therapeutic activity is concentrated in only one of the enantiomers. *Methyldopa* (Aldomet), for example, a classic treatment for high blood pressure, owes its efficacy to the S enantiomer alone.

We shall not draw out the list of drugs in which the activity of the two enantiomers is qualitatively different, for it would no doubt be wearisome to the healthy reader. Let us simply mention *ketamine* (Ketalar), whose S isomer has the desired anesthetic activity, while the R enantiomer is an excitant and can cause psychic disorders. In *propranolol* (Avlocardyl), a beta-blocker prescribed in its racemic form, the two enantiomers have markedly different properties, and so on.

In any event, in view of these various examples, one can readily understand that the question of whether a drug should be released for consumption in its racemic form or in its enantiomeric form can sometimes give rise to heated discussions. Actually, answering this question presupposes that for all the racemic compounds on the market, we first of all know the respective physiological activities of the two enantiomers that make them up. This is far from being the case. In fact, as there is no accepted international rule at present, the pharmaceutical companies themselves make the choice, based on almost exclusively economic criteria. It is obvious that a resolution process in the preparation of a synthetic active principle is an additional operation that is often difficult and always costly; placing a racemic compound on the market thus offers definite financial advantages.

Everardus J. Ariëns, of the Netherlands, is one of the "experts" who have taken the greatest interest in these delicate questions in which science and economics sometimes come up with contradictory answers. In particular, he is in the front line of the fight to have the decision whether to market racemates or pure enantiomers be subject to legal control.

If we leave aside products derived from biotechnologies and semisynthetic molecules (which derive directly from naturally chiral substances), *the vast majority of synthetic molecules used in pharmacy are marketed in their racemic form.* Ariëns furnishes us with a few interesting figures in this connection: Out of a total of 1850 compounds used throughout the world for therapeutic purposes, 523 are semisynthetic and 1327 are totally synthetic; and out of this total, 805 are achiral and inseparable and a slightly larger number (1035) have a chiral structure. Among the synthetic drugs, only 61 occur in the form of pure enantiomers, while 467 are in racemate form.

It appears, however, that things are changing. Figures have been published that provide information on these developments, as can be seen from the table below, prepared by M. Williams and G. Quallich in 1990. New products brought out on the market over the past few years (from 1983 to 1987) are divided into four groups: achiral molecules, for which the question of resolution obviously does not arise, racemic preparations, and preparations supplied in the form of a single isomer (or enantiomer), and in the last column chiral preparations of natural origin that are sold as such.

Many strong arguments have been developed by those in favor of a set of international regulations that would require commercializing only a single enantiomer recognized as the more active. The less active isomer may exhibit unpredictable and disturbing side effects, due to the fact that medical science is often ignorant of the individual reactions of patients (in the case of allergic phenomena, for example). The products of the transformation of the two components of a racemate in the organism may be different and not have the same long-term effects. Ariëns, developing these preliminary considerations, goes so far as to suggest that, in order for a user to be able to make his therapeutic choices in full awareness of the consequences, the words "Contains 50% isomeric ballast" should be imprinted on drug packages, in the same way that a smoker, when buying a pack of cigarettes, is warned of the risks involved.

	New drugs 1983–1987			*Enantio-mers*	*Natural products*
	Total	*Non-chiral*	*Race-mates*		
1983	49	19	16	11	3
1984	41	18	11	11	1
1985	40	16	13	8	3
1986	49	23	11	11	4
1987	60	16	12	25	7

These subjects, whose social and economic consequences are obviously not minor, are still being discussed. There is one point, however, on which it would no doubt be relatively easy to reach agreement among the various interested parties: When laboratory tests have shown that a racemic compound may become a valuable drug, it should be compulsory, before it is granted a marketing authorization, to determine in so far as possible the respective activities of the two optical isomers that constitute it.

Needless to say, some representatives of the pharmaceutical industry react with hesitation, if not mistrust, to such controls and requirements imposed from the top. There are many who would wish the final choice on the marketing of racemic forms or pure enantiomers to be a matter for them alone to decide, dictated on the basis of a *case by case* examination of the various possibilities which we considered at the beginning of this chapter. We have yet to hear the last word on this matter.

LIFE HAS A "DIRECTION"

Let us listen once again to what Pasteur has to say:

"If I have made myself understood, you must say to yourself: Yes, there is a profound separation between the organic kingdom and the inorganic kingdom. This line of demarcation is characterized by two facts. First of all, no inorganic or organic product has ever been synthesized that had molecular dissymmetry right from the outset. Tartric racemates are made, but racemates are resultants of symmetrical forces. It is totally mistaken to think that we are creating dissymmetry when we produce racemates. Furthermore, dissymmetry governs chemical actions that give rise to the immediate essential principles of plant life and everything, in fact, proves it. It is essential to seek to bring into play dissymmetric forces, something that is not done in our present-day laboratories. ...

"The line of demarcation of which we are speaking is not a question of pure chemistry or of obtaining any particular product: it is a question of forces. Life is dominated by dissymmetric actions, of which we sense the enveloping, cosmic existence. I even sense that all living species are primordially, in their structure, in their external forms, functions of cosmic dissymmetry. Life is the germ, and the germ is life. ...

"In this way there is introduced into physiological studies and considerations the idea of the influence of the molecular dissymmetry of natural organic products, of that major characteristic that perhaps establishes the sole clearly delineated line of demarcation that can be drawn today between the chemistry of still nature and the chemistry of living nature."

Let us disregard for the time being the vitalist connotations of these words. The questions raised by Pasteur are good questions; in other words, they are difficult and embarrassing. There is but one specific point on which chemistry today can answer him with a few limited experimental findings, for it is unable to propose any theory concerning the origin of natural molecular chirality that is both general and convincing.

It does know how, by using the action of polarized light, to *destroy* selectively one of the antipodes that make up certain racemates and bring about the appearance of a measurable rotatory power of the one that has been spared. These results were obtained in 1928-1929, by Richard Kuhn (1900-1967) *et al.* Conversely, Henri Kagan *et al.* showed in 1971, using certain photochemical reactions that bring into play polarized light in a certain direction, that from nonchiral raw materials it was possible to *synthesize* optically active *helicenes*. In the latter case, these findings are particularly spectacular due to the fact that these hydrocarbons have considerable molecular rotatory power. To be sure, in both these cases the reactions are carried out with extremely low yields and are devoid of any practical value; yet they do have the merit of demonstrating, at least theoretically, that naturally polarized light can reveal a certain molecular asymmetry.

The problems that Pasteur "sensed" a hundred years ago with so much insistence have yet to be formulated in modern-day terms. After having confined ourselves to observing that the chemical substances produced and used by living organisms are quite generally chiral, how can we explain that those products, as relates to their absolute configuration, belong just as generally to the same "series"? Let us first make this last question a little more explicit.

Amino acids are fundamental constituents of proteins, while sugars are fundamental constituents of nucleic acids (DNA or RNA). It is on them that life on earth depends. For the human species, for example, there are *eight* amino acids that are regarded as *indispensable*. They all belong to what Fischer called the L series. Thus, although they are not directly derived from one another, they are closely related sterically; three out of four of the asymmetric carbon substituents that they all include are superimposable (NH_2, CO_2H and H). The sugars found in DNA and RNA (ribose and deoxyribose) have the same absolute configuration D (still following the same nomenclatural conventions).

If we consider that the *propagation* of chirality, as we saw previously, involves chemical processes which for the most part have nothing mysterious about them, the questions posed actually relate primarily to the problem of origins. Where does the molecular chirality of natural products that seems to be closely connected with the appearance of life come from? Here again, let us try to formulate these questions in a more precise, detailed manner.

Was it an asymmetrical synthesis reaction that created the first single variety of enantiomers or is it the result of the selective destruction of the "wrong" constituent of an original racemic compound? Whichever of these possibilities we adopt, what might the dissymmetric agent that has come into play be? Was the occurrence of an optically active molecule the result of an isolated phenomenon or did it repeat itself several times? Might not the predominance of a preferred absolute configuration D or L, of which we have spoken earlier, be due simply to chance? Did asymmetry appear at an early or late stage of what some have called chemical evolution?

Obviously there can be no question here of our reexamining in detail the multiple hypotheses that chemistry—the science of matter, and a materialistic discipline *par excellence*—has formulated in order to give a plausible description of the evolution which, over several million years, led from very simple "primitive" molecules (hydrogen cyanide, for example, to mention only one) to the constituents of our brains. A few of them have been summarized in *The Chemistry of Life*, by Martin Olomucki, published in this same series. In it the reader will not fail to note, however, that the problems we are considering here were long ignored by the majority of those who wrote on the origin of life—ignored or dismissed, especially in works intended for the general public, as if the question were only of secondary importance, or nonspecialists were deemed *a priori* incapable of understanding them. This is simply one more reason for trying to bring them to light; in doing so, however, we must stress how difficult it is to have any definitive ideas in this field, even for scientists themselves. Here, more than elsewhere, modesty is essential: here the chemist enters into a domain in which the time spans involved are no longer on the scale of his or her everyday science, experimental verification is rarely conceivable, and contradictory hypotheses are disputed on the basis of controversial observations or signs. But our uncertainties must neither rule out nor discourage our curiosity.

The Swedish chemist Jakob Berzelius was the first, in 1834, to try to analyze the organic matter content of a meteorite that fell in the region of Alès, in the *département* of Gard, in order to ascertain the possibility of extraterrestrial life. But the hypothesis, suggested back in the late 18th century, of germs or organic matter from outer space landing on earth was developed especially by another Swede, the great physical chemist Svante Arrhenius (1859-1927). We shall see that these ideas still have currency.

Working along entirely different lines, the Soviet biochemist Aleksandr Ivanovich Oparin (1894-1980) and the Englishman John Haldane (1892-1964) independently suggested, sometime around the 1930s, that life, and the matter that permitted it to emerge, were not products imported from heaven, but the result of a slow chemical evolution (which may have taken from 10 million to 1 billion years) involving more and more complex transformations of molecules, the first and most simple of which may have "come to life" in the warmth of the primitive oceans.

In their initial forms, these speculations contained no mention of the origin of chirality, be it cosmic or earthly. It was only a few decades ago that scientists saw fit to return to the questions implicitly or explicitly posed by Pasteur. Today, the two major hypotheses that we have just summarized propose rival approaches and contradictory arguments and data, which will no doubt remain difficult to synthesize for a long time to come.

To the successors of Arrhenius, the "proofs" of the hypothesis of chirality of extraterrestrial origin are essentially "experimental": what must be done is to find, analyze and determine, in meteorites or in geological dust fallen from the sky, chiral organic materials which one is certain have come from elsewhere. Inasmuch as such products exist in meteorites only as traces, the possibility of contamination by substances of earthly origin is difficult to rule out and every precaution must be taken in order to eliminate such a source of error.

In 1970, R. A. Kvenvolden *et al.* reported that certain amino acids were present in a meteorite that had fallen the previous year, the Murchison meteorite. None of those isolated and identified corresponded to amino acids ordinarily found in terrestrial organisms: all were racemic (within the limits of precision of the measurements that were possible).

Within a similar perspective, the recent work of Kevin Zahnle and David Grinspoon seems to have been in line with these findings. NASA researchers have studied sediments collected at Stevns Klint, in Denmark. According to them, the sediments might be the result of the deposit of dust given off by a giant comet, at the dividing line between the Cretaceous and Tertiary periods. Here again, among the amino acids which they found there, *isovaline* (absent from the living world in the natural state) is *racemic*.

At the same time Engel, an American, and his team returned to the same subject several times (their last communication is dated 1990) and engaged in far more sophisticated analyses of the Murchison meteorite. This time, the extraterrestrial origin of the products extracted from it seem to be confirmed: the cosmic amino acids in question do indeed appear to be richer in ^{13}C isotope than their terrestrial analogues. Also—and this is a crucial observation—certain amino acids isolated from the meteorite, in particular *alanine* and *glutamic acid*, appear to be *partially enriched with L enantiomers*, and therefore belong to the series found on Earth.

As we can see, the question is still far from being definitively settled. However, even if we admit the possibility of being certain that there do in fact exist chiral extraterrestrial amino acids, the origin of the chirality of those materials will simply have been pushed back beyond the frontiers of our planet and will continue to intrigue us.

The theories that attribute to life on Earth a local origin have been far more concerned, at least more recently, with the problem of chirality, its appearance and its manifestations. The hypotheses to which they refer are of various natures. Let us try to point out the essential aspects of the most important ones.

The discovery, in 1957, of the fact that "parity" is not conserved in a certain number of natural processes earned the Chinese-born American physicists Tsung-Dao Lee and Chen Ning Yang an almost immediate Nobel Prize. Unfortunately, the importance of this theoretical advance is difficult for anyone who is not a specialist in particle physics to appreciate. It tells us, in a word, that in our universe not everything is always symmetrical. Might not this dissymmetry at the subatomic level account for the chirality that chemists or biochemists observe at their level? There are several ways to contemplate the consequences of what physicists call "parity nonconservation in weak interactions."

The first entails, in the original racemic compounds, the selective destruction by radiations of radioactive origin (*radiolysis*) of the enantiomers of the now "unfindable" series. This hypothesis, as we can see, is somewhat reminiscent of Kuhn's experiments that brought into play *photolysis* induced by ordinary polarized light and led to the asymmetric production of certain enantiomers from the corresponding racemate.

The *beta* radiation produced by radioactive strontium (^{90}Sr), retarded by appropriate means, loses part of its energy through the emission of gamma rays. The radiation thus slowed down (whence its internationally adopted German name *Bremsstrahlung* or "braked radiation") is circularly polarized to the right. Starting in 1959, certain researchers, such as Ulbricht, a German, tried to induce optical activity in molecules irradiated with this type of radiation. But it was the experimental findings published in 1968 by a Hungarian, A. S. Garay, that led to the belief, for a moment, that the mystery of the origin of chirality on Earth had at last been penetrated. After 18 months of irradiation, Garay thought he had demonstrated that D-tyrosine (an amino acid found in silk that plays an important role in the formation of thyroid hormone) was destroyed to a greater extent than the related L enantiomer subjected to the same treatment. Unfortunately, the American team of D. W. Gidlmey, working with other amino acids, was unable to confirm those findings.

A second and more theoretical approach that was also based on the existence of parity nonconservation led to the demonstration of a certain energy difference between the two enantiomers of a chiral substance—a difference which, while minimal, might have resulted, on being amplified over thousands of millennia, in the exclusive selection of the preferred one.

Among the fundamental interactions that physicists speak of, gravitation, electromagnetic forces and strong nuclear interactions have no preferred "direction"; on the other hand, the fourth, weak interaction may be at the origin of a certain dissymmetry of the chemical elements themselves and thus account for the chirality of natural organic substances.

In 1967, the unification of electromagnetic interaction and weak interaction through the work done by Steven Weinberg, Abdus Salam and Sheldon Glashow (Nobel Prize, 1979) implies not only that *all* atoms, whether free or combined, possess some optical activity, but that there exists a difference in binding energy between the enantiomeric molecules.

There would be no point here in going into the details of the calculations of the difference in energy between amino acids of the D and L series. The important thing is the result of those calculations; the balance is tipped in favor of the "natural" series. These computations, first performed by Hergstrom in 1982, were subsequently taken up again by Stephen Mason, Tranter and several other researchers. The most recent results on the subject (1990) confirm that L-alanine and L-serine are indeed theoretically more stable than their antipodes, due simply to "nonparity."

This energy predominance is obviously extraordinarily weak (a measurement of 6×10^{-19} electron volt has been given). What this ultimately means is that the "excess" that broke the equality in a racemic mixture of alanine, for example, would amount to *a single additional molecule* of L-alanine for a population of 6×10^{17} molecules (6 followed by 17 zeros).

The question that immediately arises is how this barely perceptible initial "predominance" ultimately led to near-absolute hegemony; once we have admitted this minuscule original difference, what might the possible mechanisms of its extraordinary amplification be?

Here, we return to an area which, to the chemist, is not entirely unknown. We saw earlier that a chiral reagent that has to choose between two enantiomeric molecules may react more quickly with one than with the other. This *recognition of chirality*, already encountered by Pasteur, guides us in the direction of possible solutions to the problem. Which is not to say that such solutions are obvious and controllable—far from it. While it has been demonstrated both theoretically and experimentally that the polymerization of an alanine enriched with L enantiomer, for example, preferably leads to a polymer having the same sign, this artificial polypeptide is still far from resembling a functional protein in the world of living things.

We suspect that in a field where it is so difficult to subject hypotheses to tests that prove them false, a profusion of theories is not a good sign (recall, incidentally, the ideas of the British philosopher Karl Popper). William of Ockam, the *Doctor Invincibilis* (ca. 1285-ca. 1349), made the same recommendation, known as Ockam's Razor, seven centuries ago in an age when philosophers spoke Latin: *Pluralitas non est ponenda sine necessitate* [Entities should not be multiplied unnecessarily]. Some, like V. I. Goldanski *et al.* (USSR), maintained, for example, that the chirality of the materials that make up living things was due only to chance and owed nothing to parity nonconservation, and that the molecules in question were formed and selected in the extreme cold of interstellar space.

Without a doubt, the question of the origin of the chirality of living organisms is not a simple one, and the answers to it are still less so: the difficulties which science runs up against here are on the same scale as the enormous problem posed. Yet is it not our fate as human beings, who know how to do so many things, to search tirelessly for impossible certainties?

CONCLUSION

Molecular chirality, as we have seen, has combined crystallography, optics, chemistry, and biology since long before the organizers of science added the word *multidisciplinary* to their repertory. Need it be mentioned, moreover, that this assembly of curiosities and skills was never the subject of any "call for bids" or "voluntary inducements" on the part of those who control research? It came about, one might say, without forcing talent. The field we have just visited together was opened up by available minds exercising their genius along the hazy frontiers of several disciplines: etymological borderline cases, so to speak. I am sure the young Pasteur would not be easy to fit into one of the various committees of the CNRS (French National Center for Scientific Research).

The patient and attentive reader will certainly have remarked, throughout this medley of different chapters, the number of questions raised that are still open. Let us recall a few of them:

What are, for example, the laws that govern the formation of the various species of racemates and the "stacking up" of chiral molecules? Why is it that in racemic crystals some prefer a *homochiral* arrangement (in which rights match up with rights, and lefts with lefts), while in the majority of cases these asymmetric molecules choose to associate with their "doubles?" Are there reasons, belonging to the area that mathematicians call "discrete geometry," why certain enantiomeric figures fill space better in accordance with one or the other of these possibilities?

Will it be possible in the near future to select, without laborious trial and error, from among the salts that Pasteur was the first to use, those that permit the most efficient resolution?

Will physics provide the chemist with convenient asymmetric means for practically orienting his or her synthesis at will toward the production of one or the other enantiomer?

Will we one day be able, by means of a crucial experiment, to test the various contradictory hypotheses on the stereochemical selection of the molecules that are at the origin of life?

Here let us end the list of questions that might seem to be important. Those relating to details that should also be included are hardly less exciting. Why is *aspartame* sweet? Why does a biological receptor react stereoselectively vis-à-vis a chiral agent *that it has never met before*?

The history of science tells us that the answers to questions raised—when they have finally managed to be raised—have rarely been found where they were expected or by those who were awaiting them. In any case, the important thing is that they should be found.

"In the field of observation, chance favors the prepared mind," Pasteur said in an official address delivered at the Lille Faculty of Science on December 7, 1854. Let us add another statement from the same source: "Knowing what to wonder at is the first movement of the mind toward discovery."

How, in fact, can we not be struck once again by the importance of fortuitous observation and experience in the emergence and development of the subject we have tried to render accessible in these pages? Going hand in hand with the rigorous processes of deduction for which physics furnishes us models, there exist and will no doubt continue to exist for a long time areas in which a knack and a flair, together with the unexpected, have played—and will play—an essential role. In the last analysis, it is comforting to think that not all the sciences unconditionally require heavy "equipment," of which mathematics is neither the most cumbersome nor the most costly. It may be well worth recalling this outmoded idea at a time when computerization sometimes causes one to forget the content of what is being computerized: intelligence will surely never be artificial.

Finally, the history of science has the meaning that historians wish to impart to it. They have the right to find in it whatever lessons they deem useful, provided that they do not abuse that right. At the end of this short volume, we might suggest the substance of some of these lessons: modesty and rigor vis-à-vis the facts; the proof that the best minds can sometimes be wrong; the foolishness of predictions that cannot be checked; and finally optimism regarding our capacity as humans to understand the world, slowly, laboriously.

BIBLIOGRAPHY

PASTEUR, L., VAN'T HOFF, J. H., WERNER, A., *Sur la dissymétrie moléculaire* [On molecular dissymmetry], Christian Bourgois, Paris, 1986.

MASON, S. F., "*Molecular Optical Activity and the Chiral Discriminations,*" in *International Reviews in Physical Chemistry* 23, p. 217, 1983.

MORI, KENJI, "The Synthesis of Insect Pheromones," in *The Total Synthesis of Natural Products*, vol. IV, Wiley-Interscience, 1981.

JACQUES, J., COLLET, A., and WILEN, S., *Racemates, Enantiomers, Resolutions*, Wiley, 1981.

JACQUES, J., *Berthelot, autopsie d'un mythe* [Berthelot, the autopsy of a myth], Paris, Belin, 1987.

ROSNAY, J. DE, *L'Aventure du vivant* [The adventure of living things], Le Seuil, Paris, 1988.

GARTNER, M., *The Ambidextrous Universe*, Penguin Books, 1982.

OLOMUCKI, M., *The Chemistry of Life*, McGraw-Hill, New York, 1993.

BALIBAR, F., *The Science of Crystals*, McGraw-Hill, New York, 1993.

SCHATZMAN, E., *Our Expanding Universe*, McGraw-Hill, New York, 1992.